Integration and Visualization of Gene Selection and Gene Regulatory Networks for Cancer Genome

Integration and Visualization of Gene Selection and Gene Regulatory Networks for Cancer Genome

SHRUTI MISHRA, PhD
Associate Professor
Department of Computer Science & Engineering
Vignana Bharathi Institute of Technology (VBIT)
Hyderabad, India

DEBAHUTI MISHRA, PhD
Professor
Department of Computer Science & Engineering
Institute of Technical Education and Research
Siksha 'O' Anusandhan (Deemed-to-be University)
Bhubaneswar, Odisha, India

SANDEEP KUMAR SATAPATHY, PhD
Associate Professor
Department of Computer Science & Engineering
Vignana Bharathi Institute of Technology (VBIT)
Hyderabad, India

ELSEVIER

ELSEVIER

3251 Riverport Lane
St. Louis, Missouri 63043

INTEGRATION AND VISUALIZATION OF GENE SELECTION AND
GENE REGULATORY NETWORKS FOR CENCER GENOME   ISBN: 9780128163566

Content Strategist: Rafael Teixeira
Content Development Manager: Kathy Padilla
Content Development Specialist: Pat Gonzalez
Publishing Services Manager: Deepthi Unni
Project Manager: Nadhiya Sekar
Designer: Gopalakrishnan Venkatraman

Printed in United States of America

Last digit is the print number: 9 8 7 6 5 4 3 2 1

Contents

CHAPTER 1

Introduction

SHRUTI MISHRA, PhD

OVERVIEW

Data mining and *bioinformatics* have played a vital and key role in the development of system biology. Preferably, in most of the cases they work hand-in-hand. Researchers have paid a huge attention toward the growth of system biology and molecular biology. These areas have been determined from the subject of the individual gene from the huge gene expression information.[1−4] DNA microarray is one such technology that permits to evaluate the expression levels of thousands of genes using a single experiment. These expression levels are generally useful in diagnosis or classification of the disease-causing tumorous genes.

The advent of microarray technology aims at discovering the mRNA levels of the gene expression data that can be further used in many diversified applications. The applications that are functions of these microarray technology discoveries are drug discovery and understanding its response in different organisms, understanding of biologic knowledge and pattern discovery, getting hold of clusters and subclusters of diseases, classification of patients with diseased and nondiseased group, and many others.[5−7] However, a major issue that DNA microarray data have always traded with is high dimensionality. Each individual data are expressed with more than thousands of expression levels. Hence, the data exploration or classification task becomes really tedious in high-dimensional datasets because of the factor of curse of dimensionality. To distinguish the same more further, it is now important to realize that cancer classification is one of the most critical research topics in molecular and systems biology.[8,9] The study of this disease in the biomedical research allows us to obtain a more detailed insight about the gene expression levels and patterns of the genes under different circumstances or weather.

As said earlier, the task of cancer classification is rather moved by the fact that there is usually more number of genes but less number of sample sizes. Hence, a skeptical feeling does pass off when testing

an experiment for its predictive classification accuracy. Likewise, biologic systems are normally viewed as a repository, which mostly contain genes that are a kind of DNA.[10,11] Now this gene information gets transcripted into RNA, which is then transformed into proteins. Gene profiling is also a major attraction for the classification of cancer disease.[12,13] It permits identification of the tumorous classes, which can be further processed with certain drugs for the development of more appropriate handling for different individuals.

GENE REGULATORY NETWORK AND GENE REGULATORY NETWORK VISUALIZATION

As discussed in the previous section, it can be inferred that genes are the vital building block of life. Not only the genes but also the products that they yield are an important part in treating life. Still, for the cell to function properly, it is quite important and relevant that the genes should interact with each other.[14−16] This kind of interaction usually forms a huge computational network within the arrangement, which is later tagged as gene networks.

Gene networks are usually formed when genes interact among themselves in certain conditions. These gene networks depict the relationship between the gene sets. In other words, these gene networks are also known as gene regulatory network (GRN). The GRN can be technically described as an aggregation of DNA segments in a cell where a heavy interaction takes place among these segments (directly or indirectly), by governing the overall rate at which genes are transcribed into RNA.[17,18] Gene networks or GRN is well suited for the qualitative and quantitative modeling and simulation as desired by scientists and researchers. These networks can be modeled using various techniques and methodologies.[19,20] Some of them are *correlation analysis, Bayesian networks, Boolean networks, dynamic Bayesian networks,* and *neural networks.* GRNs have their own basic applications like they allow us to infer and derive the biologic hypothesis occurring in a living organism. In other words, they allow researchers to understand the transcriptional factors that are involved in regulating the genes. Usually, statistical methods are most suited in understanding the molecular patterns and interactions of GRNs.

Visualization of GRN is an important aspect of molecular biology. Owning a huge gene list with thousands of expression levels does not mean anything till any statistical inference has been sucked out of it. Applying a statistical technique, a theory can be chalked out and gene-gene interaction list can be furnished. With this gene-gene interaction details, GRN network can be constructed and visualized using any software tool available. Here, Cytoscape has been used, as it is regarded as the simplest and most well-understood tool for the visualization of GRN.

GENE SELECTION

Gene selection[21–23] is a primary step for the construction and visualization of GRN. As we know, the gene expression set is huge and thousands of genes do occur over there with many expression levels measured under different circumstances. Only one of the questions that arise is, do all of them have significant role? The result is simply no. All genes do not have any part in causing diseases. Rather, there are few genes that are far more responsible in causing disease, and they behave differently in different conditions. Technically defining, gene selection is a technique that is widely used in data mining where a small subset of genes are selected from the huge list of genes. The selected genes are considerably thought as relevant to the nature of the problem. The genes selected usually provide a greater impact on the quality of the model that is selected that best defines the given problem. Hence, the interactions and study of these interacting genes are important, which can be ultimately achieved through the construction and visualization of GRNs.

From the biologists' and scientists' point of view, gene selection is a basic step through which classification accuracy can be measured. Using all genes and predicting classification accuracy from the same usually lead to the incorrect decision accuracy prediction. Hence, it is really important to identify the small subset of genes that would ultimately lead to correct decision making. In literature, many gene selection algorithms do exist that are needed to choose a subset of genes, using which higher classification accuracy can be obtained.[24–26] Some of the traditional gene selection techniques are filter method, wrapper method, embedded method, and hybrid method. Filter method

is applied to evaluate each feature separately. These are commonly practiced in high-dimensional datasets, and here the method is classifier independent. The wrapper method uses classifier performance as the feature or gene evaluation criteria. It is commonly viewed as a stochastic and deterministic method. The embedded method considers the use of model representation and its properties for analysis of the problem and feature selection. Finally, hybrid methods are the ones that combine two popular feature selection methods and practice the combined method for feature selection.[27] It has been commonly found that the methods of gene or feature selection are fairly similar to each other. What really varies is the model of selecting the genes of the large gene expression dataset. There is a statistical technique that allows selecting genes using the concept of ranking. Here, the genes have traditionally been ranked using various ranking algorithms, and then from the sorted, ranked gene list, top 50, 100, or 150 and more genes are selected. This is a considerable method that can be used on gene selection, as statistical inference becomes one of the selection criteria.

MOTIVATIONS

Some of the major motivations that diverted us for holding up this research work are expressed in the following:

- *Gene selection and classification accuracy*: One of the major reasons for undertaking this area of research is to find a small subset of genes responsible for cancer, based on which classification accuracy can be taken up and correct decisions can be made. Using all genes for discriminating cancer and classifying cancer basically lead to incorrect conclusions.
- *Visualization of GRN*: The genes selected from the perspective algorithm can be further created into a network forum where the connection and interactions of each gene can be assessed. The one that would be aloof would definitely be not getting affected by the activities taking place in other genes.
- *Understanding of behavioral pattern*: Creation of the network also allows a flexibility of understanding about various oncogenes, their expression strategies, and behavioral pattern. The interaction between the genes provides us with an opportunity to understand the relationship taking place among the interconnecting genes.
- *Application specific*: This study of behavioral pattern and oncogene leads us toward information that can be shared for better drug discovery. However, there are many other applications that can be extracted, but drug discovery remains the vital one.

CHALLENGES

Few basic challenges have been taken up and have been resolved in this domain. They are specifically mentioned, and corresponding solution has been provided for each. These challenges are listed beneath:

- *Gene selection*: As stated earlier, all genes do not participate in the occurrence of disease. There are only a few genes that are highly disease causing and get affected easily by the environment or conditions and surroundings. Hence, it is important to identify a small subset of genes that can help in decision making and for diagnostic purpose in clinical practice. For this, three different genres of gene selection algorithms were proposed and discussed with different types of datasets. The proposed gene selection algorithms are SVM-BT-RFE (support vector machine-Bayesian *t*-test-recursive feature elimination; it was found to be powerful in finding the minimum subset of genes that further helped in attaining the maximum classification accuracy), CCA-TR (canonical correlation analysis-trace ratio; it does not have any dependency factors unlike SVM-BT-RFE, where weight vector generation was a major requirement), SNR-TR (signal-to-noise ratio-trace ratio; it does not involve hefty calculation unlike CCA algorithm for adjacency matrix calculation). Altogether the three methods proposed vary in different aspects such as time, efficiency, and computational complexity.
- *Validation of the genes selected*: Now, as a small subset of genes are selected, they need to be validated for the satisfying motive behind their selection. As we know, the gene list is huge and finding a suitable set of handful genes is really tedious and difficult. Hence, the selected genes have been validated using different classification accuracy for finding their predictive performance value. Different classification algorithms are expressed in the literature, of which only few specific algorithms are applied.
- *Improvised methods for further validating the genes:* Classification is a generalized and popular way of validating a specific lot of genes discovered. Survey says there exist many other methods that can be explored for validating the selected set of genes. Some of them are the probabilistic measures such as Kuncheva's stability index (KSI), balanced classification rate (BCR), and balanced error rate (BER), *G-Mean*, and *F-Score*.[28] In this study, the first three measures are used for further exploring the validity of the selected sets used.
- *Visualization of GRN*: Selection is one of the primary steps for visualization of the genes in the form of a network. The form of network construction that we are using is called *reverse engineering process for the visualization of GRN*. The usual process says that using the whole gene expression dataset, the relationship among the genes can be created and the network would be made usable. After that from the huge network, a small subset of genes can be selected based on some criterion measure. However, this method is rather complex. Hence, a *reverse engineering approach*[29] has been used in this domain. Network discovery always allows a schematic view of how the genes interact and how they get connected with each other. However, a frequent case that has been observed is that all genes are not some or other way connected. In other words, there exist some genes that usually are disconnected from the rest of the set. An inference can be made out of the same, where it can be reasoned that those genes that are disconnected from the generalized network do not really affect the other genes under different extreme conditions.

ORIGINAL CONTRIBUTIONS

Here, in this research, three different genres of gene selection algorithm have been proposed. Although there exist many gene selection algorithms, our proposed algorithms have been extraordinarily performing well. SVM-BT-RFE, CCA-TR, and SNR-TR are the three gene selection algorithms that have been proposed and experimentally evaluated in this study. In each case a subset of genes were extracted from the huge gene expression datasets. Colon cancer, leukemia, medulloblastoma, lymphoma, and prostate cancer datasets were used for experiments satisfying the correctness and efficiency of the gene selection algorithm.

SVM-BT-RFE was proposed as an extension of SVM-RFE and SVM-T-RFE. Here, a new ranking method was introduced, which removed the time, complexity, and unnecessary iterations that were getting undertaken by the latter two methods. In CCA-TR, trace ratio (TR) algorithm was exploited into a deeper segment and a new scoring method was introduced to it. As we know, the TR algorithm uses either Fisher score or Laplacian score for finding the adjacency matrix calculation and trace calculation. A CCA technique (statistical technique) was used as a replacement for the generalized Fisher score or Laplacian score, and based on this, the new adjacency matrix set and trace was generated. Here, it has been found that the CCA-TR involved a lot of hefty calculation of variance and covariance. So, we thought of integrating a new technique with the TR algorithm. The SNR was used for this purpose. It is one of the most elementary and easy-going techniques for detecting the gene ranks. The same was used in place of CCA or Fisher score or Laplacian score for training a new scoring equation for the generation of trace parameter. This was quite simple, as it did not involve any complex calculation unlike CCA-TR.

All these gene selection techniques were followed out on the five chosen datasets with a varying dimensionality. Finally, the selected genes were validated and tested for their predictive power. For this, classification method or algorithms were used such as SVM, variants of neural network (NN) such as *resilient propagation, backpropagation, Manhattan propagation, and quick propagation*. Although more classification techniques could have been used, our study was limited to these techniques. Classification accuracy cannot be termed as the sole method available for validating the genes selected. In fact, many techniques do exist for validating the same. Some of the popular probability measures such as KSI, BCR, and BER are used in this work for enhanced validation of the genes selected.

Ultimately, we aimed at visualizing the relationships that take place among these selected genes. For the same, the GRN was used for the construction and Cytoscape tool was used for the visualization of the network. Pearson's correlation coefficient was the statistical technique that was applied for bringing forth the relationship among the selected genes. The visualization of the network provides us with a scope to realize which genes are really interacting with each other and how they are behaving under different circumstances. It is recognized that, under different environmental conditions, genes behave differently, and this is the reason why one gene's effect is transferred to the remaining genes connected to it. In addition, from the visualized network, it can be clearly inferred which are the genes that do not connect to the standard network, that is, which genes stay aloof and are not a part of the entire network. By this, it can be concluded that, although they contribute to the classification accuracy, they are really not involved in affecting the other genes connected to the network.

BOOK ORGANIZATION

The research work has been properly divided and stated as follows. In this chapter, a thorough overview and focus toward the research study is discussed. Other than that, the fundamental techniques along with their usefulness are also discussed herewith. Motivations and some of the research challenges that have been carried out are also acknowledged in this study. In Chapter 2,

a detailed literature survey and background stories of the major techniques are discussed. Major techniques such as GRN, gene selection, information gain (IG), TR, CCA, and many are more clearly defined. Many authors have contributed a lot toward these fields and aspects; all their work and contributions are made available in the chapter.

In Chapter 3, our focus shifts to our proposed gene selection algorithm. SVM-BT-RFE is the proposed gene selection algorithm that was used as an extended and modified version of the existing SVM-RFE and SVM-T-RFE. Providing and proposing a new modified rank is a major goal in all the three techniques. The ranking parameter used in the latter two methods had their own specific issues such as removing one gene at a time, time complexity, and threshold dependency factor. In SVM-BT-RFE, all these issues were categorically addressed and resolved. Other than that, the number of genes selected provided a better performance accuracy with SVM classifier when validated for their authenticity. In Chapter 4, two gene selection algorithms were proposed, that is, IG-TR and CCA-TR algorithms, which are said to remove the basic issue of the previously proposed SVM-BT-RFE algorithm. The SVM-BT-RFE algorithm heavily depends on the weight vector calculation by using the SVM classifier. The entire ranking algorithm depends on this. However, our proposed method is independent of the same. Both the algorithms were into diversified directions. In IG-TR, IG was the main base of the algorithm. Here, the IG of all the genes was extracted and the same was used for reordering and sorting the dataset. Now, this sorted dataset was passed to original and unmodified TR algorithm. The results produced by selecting top-ordered 50, 100, and 150 genes were undoubtedly good, but the major problem that persisted throughout was that finding the IG value and then sorting the huge ordered dataset were more complicated. Hence, we tried and modified the TR algorithm instead of modifying the data input. CCA being a statistical technique would usually provide a better result. By thinking the same, we replaced the usual Fisher score rank with CCA score rank for generating the adjacency matrix and trace value. Thus, when passed to the different classifier, it produced a satisfying result with the same top 50, 100, and 150 genes that were selected. To further validate the genes liability, we passed the same to the three performance indexes or probabilistic measures, i.e., KSI, BCR, and BER, where again it produced a satisfying result.

In Chapter 5, another gene selection algorithm was proposed using the concept of TR algorithm. In the previous work, it is no doubt that TR algorithm produced a satisfying result, but it had its own demerits. One such demerit is the tedious and time-taking calculation involved in it. CCA needs a heavy calculation of variance and covariance for a matrix of large dimension. Hence, we thought of replacing it with a simplified and modified scoring technique called SNR, which is the simplest technique that uses a calculation of mean and variance. This modified algorithm was then given to the five datasets, and its accuracy was assessed. In most of the cases, it was observed that it performed well than the existing original TR algorithm with a less number of iterations. Other than that, the same three probabilistic measures were used for further reassuring the validity of the genes. In all the cases, the results produced were far more convincing. In Chapter 6, a GRN visualization of all the genes selected from the three proposed techniques was made. In other words, the top-ranked 50, 100, and 150 genes derived from the three gene selection algorithms were presented in a structure with some validating relationship. This relationship among the genes was established using Pearson's correlation coefficient, which itself is another statistical technique.

A threshold parameter of 0.6 (derived from the literature) is used for the construction and visualization of the network. Cytoscape tool was used for the design of visualization of the network.

Finally, in Chapter 7, a proper conclusion or summary of the above work is illustrated with a scope of providing light to the future.

SUMMARY

In this chapter, a distinct discussion about the research work was carried out. Other than that, some of the fundamental techniques such as GRN and gene selection were also discussed herewith. For initiating this research, there were many factors that motivated us such as extracting information for the analysis of oncogenes. This chapter also speaks about the major challenges that have been undertaken for our work along with the brief view of how those challenges have been addressed. In the next following chapter, a detailed background study and literature survey of the major techniques are given. Many authors have contributed to this related field; hence a brief overview of the same is depicted and represented.

CHAPTER 2

Literature Review

SHRUTI MISHRA, PhD

BACKGROUND STUDY

This section works out with the background study and research contributions of various researchers and authors. The detailed insights about the methodologies are discussed herewith.

Gene Regulatory Network

In this new age, one of the major problems addressed by the system biologist is the process of understanding gene regulation in different segments of biologic environments. GRNs have always played an important synthetic role in the discovery of molecular interactions and patterns in system biology. The GRNs can be clearly inferred from the gene expression profiles in the form of interaction or pattern graphs. In fact, visualization of these networks forms a vital part in finding the relationships among the genes present in the network.

Ironi et al.[30] many aspects in the algorithm and implemented things using mathematical methods for the study of the dynamics of GRNs. They proposed a computational tool known as GRENS for simulating the dynamics of GRN models. It was found to be quite efficient and it well handled the constraints and their specifications.

Chen et al.[31] proposed a learning algorithm called *fuzzy cognitive map learning algorithms* for the discovery of regulatory networks from the gene expression data. They used the error as the objective function, and based on this, the feasibility of the approach was assessed. The datasets used for their domain were based on the time series data, a good correlation was found between the simulation error and model error. Montes et al.[32] tried to discover the dynamics of regulatory networks by finding the patterns that lead to the formation of the fully developed gynoecium. In other words, they developed and reconstructed a gynoecium regulatory network that allowed an understanding of the molecular interactions for finding the gynoecium cell identities.

Patel et al.[33] tried to grow an integrated database through which all the required biologic processing can be performed in single platform. *Leishmania major* and

Schistosoma mansoni databases were proposed, which integrate the biologic networks and regulatory pathways for determining the facts and information about the genome sequences for different organisms. The database also helped in receiving a copy, where information about the rule of the lipid metabolism can be mixed with the transcriptional factors, regulatory genes, and proteins for operating the genetic levels. Chowdhury et al.[34] developed a method for modeling large-scale GRNs by decomposing the given GRN into two independent subgraphs. They use biologic traits and characteristics of GRNs for decomposing the network into two subnetworks. Aside from this, they also projected and used a cardinality fitness function to accelerate up the inference process by incorporating the maximum indegree concept. The concept of gene clustering and gene ranking had been used for the purpose of acquiring a local search method. This was required to successfully analyze the large-scale genetic networks.

A new model for analyzing the GRN was established by Ito et al.,[35] which used the verification technique. The behavior and pattern of the network were studied by the linear temporal logic where the conduct and the biologic property were qualitatively analyzed. They also offered two more methods: model checking method and approximate analysis method. The model checking method basically concentrated on separating the network into subnetworks, going over them individually, and integrating them. The approximate analysis method focused on deriving simple and gentle rule for understanding the potential behavior of the networks. Hecker et al.[36] proposed a reconstruction method for GRNs using computational techniques. The modeling techniques used by them focused specifically on the biologic data and molecular information.

In that respect are several computational methods for constructing a GRN, which have a specific use in the systems biology. Politano et al.[37] discussed how miRNAs and posttranscriptional regulatory interactions were modeled, resorting to *Boolean network*. They proposed a Boolean network based on the posttranscriptional model that implemented and simulated a Boolean network to compute the attractors of the network, taking into account posttranscriptional activities. The set of attractors of two biologically confirmed networks, focusing on the regulatory role of miR-7, was also discussed. In fact, the attractors were compared with the networks in which the miRNA was removed. As the purpose of the miRNA is to increase the network stability, in both the networks, it was highlighted, confirming the cooperative stabilizing role of miR-7.

Chueh et al.[38] proposed a pioneering method to reconstruct time delay Boolean network as a way for exploring biologic pathways. They generalized the Boolean network model for coping with the dependencies that had two sorts of relationships: similarity and prerequisite. In fact, they also suggested an approach for reconstruction of genetic network inference from gene expression data that relied on the supposition that the expression of a gene is likely to be inhibited by a relatively small number, n of genes. Kim et al.[39] introduced a variable selection process based on chi-square test (CST) called as CST-BN that reduces Boolean network calculation time and also obtains optimal network structure by using chi-square statistics for examining the independence in contingency tables. The projected CST-based boolean network (BN) adopted the best-fit extension problem to effectively regulate all possible Boolean functions. It was found that the CST-based BN method was about 6.9 times quicker than the original BN method. If the network had a generously proportioned number of nodes, then the difference between the computing times of the two algorithms would have been considerably higher.

Silvescu et al.[40] presented a generalization of the Boolean network model to address the dependencies among genes, which covers for more than one unit of time. The model called the *temporal Boolean network* allowed the expression of each gene to be guarded by a Boolean function of the expression levels of at most k genes at times. They also staged a popular machine learning algorithm for decision tree induction for inference of the temporal Boolean network from gene expression data. Liu et al.[41] proposed a two-stage structure learning algorithm that integrates the immune evolution algorithm for constructing a *Bayesian network*. The results obtained showed that the proposed algorithm was quite effective enough to learn many of the known real regulatory relationships with a high validity and accuracy. The algorithm optimized the network parameters by using the immune evolutionary algorithm and it simplified the traditional three-stage algorithm.

Adabor et al.[42] offered a search method that was a hybridized version of simulated annealing (SA) with a greedy algorithm (GA). It explored the whole search space by running through a two-phase search: first, with a simulated annealing search and then with a greedy search. The method evolved to a near-optimal solution within a fixed time without affecting the quality of the true regulatory network that was accomplished. Only a prior domain knowledge was the basic requirement for directing SAGA to achieve the results with biologic significance and usefulness. Aside from this, it did not restrict the number of nodes for effective structure learning, therefore furnishing an equal probability to any given variable to mention other variables.

Heijden et al.[43] proposed an algorithm to learn temporal probabilistic models from the clinical time series data having missing values. They anticipated a variety of learning algorithms that were based on the *naïve Bayesian networks* having attractive properties of reduced computational complexity and good prediction performance. These proposed methods were also used to build predictive models of health status of patients with chronic obstructive pulmonary disease and where it was expected that the models can manage both the dynamic nature and inherent uncertainty of the disease progression. Likewise, Wang et al.[44] suggested a hybrid constraint-based scoring-searching method for discovering gene networks from DNA microarray data. An algorithm was used to generate a Bayesian network based on dependency analysis, and the resulting construction was used to search for a scoring metric.

Labatut et al.[45] aimed to understand how large-scale network activation is extracted from cerebral information processing mechanisms that explain the conflicting activation data. Here, a formal technique based on *dynamic Bayesian network* was introduced that integrates the biologic plausibility in the framework. Kim et al.[46,47] extended the Bayesian network and the dynamic Bayesian model, which can construct cyclic regulations for the time series data. They optimized the structure of the network, which attains the best possible form of representation in the gene interactions described by the data with noise. Their model was capable of considering time information into account and can analyze the microarray data as the continuous data without the extra data treatments such as discretization. Even the model is capable of extracting the nonlinear relations.

Chen et al.[48] proposed a Bayesian approach for discovery of longitudinal morphologic changes in the human brain. The technique used a dynamic Bayesian network for representing interregional dependencies. Here, the dynamic Bayesian network modeling represents complex interactions among temporal processes, and the approach was validated by analyzing a simulated atrophy study. Subsequently, it was found that this approach takes only a modest number of samples to detect the ground-truth temporal model. Peña et al. [49] studied the cross-validation procedure for scoring criterion and also studied the Bayesian scoring criterion for learning the dynamic Bayesian network model. They experimentally proved that cross-validation leads to a model and it generalizes a range of sample sizes than the Bayesian network model.

GENE SELECTION

Gene selection is a primary and fundamental step in the visualization and construction of GRNs. As we know there are many genes present in the data matrix; however, all of them are not specifically responsible for disease causing. Preferably, a selected amount of genes is responsible for disease causing. The discovery of these specific genes is causing a major attraction in the world of biologic and molecular research where identification of specific genes and their behavior pattern can help in disease diagnosis and drug discovery. Other than gene selection, gene ranking is also an important factor of consideration for which different methods are available in the literature for study of class data. Some of them are fold change, moderated *t*-statistics, and significance analysis of microarrays (SAMs). There is another method called as rank product (RP) method that is the only rank-based nonparametric method. This method independently handles upregulated and downregulated genes under one class and therefore produces two separate ranked gene lists.

Many research studies have been focused and related to this topic of gene selection and gene ranking. Some of them are discussed herewith along with some existing traditional methods that are used for the determination of gene selection. Andrusiewicz et al. [50] used a reverse arrangement, followed by quantitative polymerase chain reaction (RT-qPCR) method for interpreting the construction level of seven genes that can be treated as reliable reference genes. Of these seven genes, they selected two genes to be used in RT-qPCR method for which they made a thorough study and analysis. Mohammadi et al. [51] anticipated a maximum-minimum correntropy criterion (MMCC) approach for the selection of significant genes from gene expression datasets. They also used an evolutionary optimization technique for the purpose of finding optimality from the number of significant genes selected. The selected informative genes were further validated by using support vector machine (SVM) classifier, which provided a maximum accuracy level for the gene expression sets considered. Aguas et al. [52] suggested how the feature or gene selection methods can be used

for achieving maximum reliability in classification of viral sequences by different species in genomic data. Man Chon et al. [53] proposed a novel feature or gene selection technique that was used with a multiple classifier method for identification of cancer-related mutations in cancer datasets.

Fuentes et al. [54] made an investigation where stability of eight reference genes was tested for *Solanum lycopersicum* using qPCR normalization. In other words, all the eight reference genes considered were ranked properly based on their expression stability. Diaz-Uriarte and de Andres [55] discussed a method for gene selection and data classification based on a random forest where the method yields small sets of genes, which provides high classification accuracy. Shreem et al. [56] proposed an approach that embeds the Markov blanket with the harmony search algorithm for gene selection. Cai et al. [57] too proposed a feature weighting algorithm for gene selection called local linear hyperplane-relief (LHR), which estimates the feature weights through local approximation based on ReliefF. Han et al. [58] proposed and suggested the gene-to-class sensitivity subjugated by a single hidden layer feedforward neural network in a hybrid gene selection. They used *k*-means clustering and binary particle swarm optimization for filtering irrelevant genes.

Model et al. [59] established how phenotypic classes can be predicted by amalgamating feature selection methods and discriminant analysis for methylation pattern–based discrimination between acute lymphoblastic leukemia and acute myeloid leukemia. They used SVM to the methylation data for using every CpG position as a separate dimension. Li et al. [60] examined the problem of edifying the multiclass classifier for tissue classification based on gene expression datasets. It was established that the results are good for datasets with a small number of classes and the accuracy is moderately less for datasets with a large number of classes. Mundra and Rajapakse [61] used the famed *t*-statistics for gene ranking in the analysis of microarray data. Here, they have divided the *t*-statistics into two parts: relevant and irrelevant data points. A backward elimination–based iterative approach was designed to rank genes using only the relevant sample points and *t*-statistics. It was found that the proposed method performed considerably better than the standard *t*-statistic approach. Kira et al. [62] partitioned the information points into clusters using *k-d* tree and chose random data point from each cluster, and then performed feature selection by means of ReliefF, which looks for frontier points to estimate feature weights. Pechenizkiy et al. [63] used the principal component analysis (PCA) for dimensionality reduction after partitioning large datasets with k-d tree. Cavill et al. projected a genetic algorithm (GA)/k-NN-based move for concurrent feature and sample selection from metabolic profiling data.[64]

Cawley et al.[65] proposed a straightforward Bayesian approach, which gets rid of the regularization parameter fully, by integrating it out systematically using an uninformative Jeffrey's prior. The anticipated algorithm (BLogReg) uses two or three orders of magnitude faster than the original algorithm, as there is no longer a necessity for a model selection step. Two new dimensionality reduction techniques were proposed by Fitzgerald et al.[66] These methods use the minimum and maximum information models. These are information theoretic extensions of spike-triggered covariance with the intention that can be practiced with non-Gaussian stimulus distributions to locate relevant linear subspaces of random dimensionality. Piao et al.[67] projected an ensemble correlation–based gene selection algorithm based on symmetrical indecision and SVM. In the method, symmetrical indecision was used to analyze the importance of the genes, and the diverse preparatory points of the pertinent subset were used to produce the gene subsets where SVM was used as an assessment criterion of the wrap.

Han et al.[68] proposed an extended impoverished gene selection measure based on *binary particle swarm optimization* and *gene-to-class sensitivity* information. The selected features derived from the above technique were passed to *extreme learning machine*, SVM and k-NN classifier for achieving the prediction accuracy. It was ultimately found that the result produced from the three provided a beneficial result for the selected features. Meng et al.[69] proposed a neighborhood system for combining gene expression data with biologic knowledge. They have also used rough set concept for designing a gene selection method framework. The outcomes derived showed an explicit result for the above proposed gene selection method when applied to different datasets. Liao et al.[70] suggested a gene selection algorithm based on the *locality-sensitive Laplacian score*. Here, information was passed through a geometrical shape where the within-the-class information is minimized and between-the-class information is maximized. Pang et al.[71] developed a new gene selection method using the basic concept of a random forest for identifying a set of informative genes. The proposed method was compared with several machine learning methods, and it was observed that the results produced from this method were quite convincing than the rest.

Ji et al.[72] suggested a *partial least squares*–based gene selection method for identifying tumor-related genes. The proposed method was also found to be suitable

for multiclass classification. The outcomes showed that when applied to diversified datasets, the outcome extracted from this method was rather relevant and effective. Li et al.[73] proposed an *optimal search*–based gene selection method that helped in identifying the optimal gene list. Tabu search was also used for the same, and it was found that the results produced were quite effective with respect to performance as an indicator. Bontempi[74] proposed a blocking strategy that was implemented by the forward selection process for identification of marker genes. The forward selection method used produces many gene subsets. Hence, it was important to select a specified gene subset out of all. Blocking concept significantly helped in taking away the extra gene subsets, and the same was used for bettering the performance of the datasets. Au et al.[75] proposed an attribute clustering measure that creates a subgroup or small clusters of attributed, which are quite interdependent on each other. Established on the criterion function, these interdependence clusters are selected and the performance was judged based on the classifiers available.

Support Vector Machine-Recursive Feature Elimination–Based Gene Selection Algorithm

Guoyan et al.[76] proposed a feature elimination technique using SVM known as support vector machine-recursive feature elimination (SVM-RFE). In this algorithm, the genes are removed recursively based on the SVM classifier weights and later the samples are classified with SVM.

Li et al.[77] proposed support vector machine-t-statistics-recursive feature elimination, a gene selection algorithm that extended the SVM-RFE algorithm by incorporating the Welch t-test. This method combined the statistical Welch t-test to predict higher accuracy and more significant genes. Hidalgo-Munoz et al.[78] used SVM-RFE upon EEG signals for detecting the important scalp region. For this, SVM-RFE was specifically used for detecting the scalp and identifying the relevant time interval. It is said to be outperforming the classical methods of scalp generated. Other than the usual microarray datasets, SVM-RFE was applied on infrared data too. Spetale et al.[79] used infrared data for which SVM-RFE was used for the purpose of dimensionality reduction. Yet another area of application of SVM-RFE has been stated by Shieh et al.[80] They proposed a product form feature based on consumer preference that would be highly needed for product designing. They used a multiclass-based SVM-RFE algorithm for identifying the important product form feature. This multiclass

SVM-RFE algorithm, where *Gaussian kernel* function has been used, provided good result as compared with other techniques.

Huang et al.[81] used the SVM-RFE algorithm for predicting the classification accuracy of multiclass problem on datasets such as Dermatology and Zoo. Apart from this, they also used a Taguchi parameter along with the SVM classifier for optimizing certain parameters. Ultimately, the result produced by combining the SVM-RFE and Taguchi optimization methods was drastically higher. Tang et al.[82] proposed an RFE algorithm based on *fuzzy C-means (FCM) clustering* concept where the same has been merged by SVM-RFE method. The new gene selection method that they proposed was called as FCM-SVM-RFE. Here, based on the model, each cluster-induced space was formed and genes contributing toward it were unconditionally selected. Duan et al.[83] changed the usual SVM-RFE in a different extent. They proposed a backward elimination method for the SVM-RFE algorithm for gene expression data, which was later known as an multiclass — support vector machine-recursive feature elimination (M-SVM-RFE) model. This model undoubtedly produced good gene subsets that ultimately improved the classification accuracy. The same was further corroborated by the concept of gene ontology where similarity between the pair of genes is counted and the performance for the same is assessed.

Srinivasan et al.[84] made a study of two bioinformatics applications known as HMMER sequence alignment and SVM-RFE for gene expression analysis on Intel $\times 86$ multiprocessor system. In that study, it was observed that HMMER is quite computationally sensitive and SVM-RFE was memory intensive. Hence, it exhibited a better result performance-wise in the machine by improving the memory bandwidth. In the end, it was resolved that in both the algorithms, compiler and run-time optimization measurement played a more serious part in accomplishing the best execution that was wanted. Yuan et al.[85] made a massive improvement in the SVM-RFE that is said to speed up the entire execution process of the same. They proposed an M-SVM-RFE, which is said to remove the retraining process of SVM-RFE after each iteration of elimination for calculation of new weight vectors. This concept was further passed to two benchmark datasets where it was proved that the result produced was much satisfying than the general SVM-RFE. Tang et al.[86] also proposed a modified SVM-RFE algorithm using granularity concept. It was basically a hybridization of statistical learning theory and granular computing theory for the selection of genes. The same was later known as a GSVM-RFE algorithm for gene or feature selection.

Yoon et al.[87] proposed a multiple SVM-RFE model for classification of *the mammogram*. Usually, artificial intelligence methods are applied for diagnosing the mammograms, of which feature selection is the basic criterion. For this, a modified and classifier-specific feature selection method was developed that aimed at eliminating irrelevant features found in the dataset. Wang et al.[88] proposed a modified feature selection technique for the prediction of subhealth state. For this purpose, they modified SVM-RFE for the classification problem with features. The purpose was not to simply reject the features but to retain the important features from elimination. It was observed for the said data the performance accuracy was tentatively high as compared with the generalized SVM-RFE method. Wang et al.[89] proposed an SVM-RFE for multi-SVM classifier where class interval was kept as the evaluation criterion and features were eliminated recursively. Other than this, a rare chaos particle swarm optimization was used for the classification experiments based on the gene datasets. Li et al.[90] proposed a modified SVM-RFE method for imbalanced classification of data along with an automatic filtering algorithm. The filtering algorithm was applied to generate filtering rules for obtaining higher classification accuracy. Zhang et al.[91] proposed a correlation-based filter with SVM-RFE algorithm for feature selection. It aimed at combining the best features in a multiple of groups. Extensive studies were made in the proposed algorithm, and the same was implemented on many gene expression sets, for which the results produced were much better as compared with the original SVM-RFE. Besides, the predictive classification accuracy was found to be dependable for the above said datasets. Yin et al.[92] proposed a correlation coefficient−based method for removing the redundant data. After removal of these redundant data, they thought of using SVM-RFE to remove the irrelevant features, as SVM-RFE is only capable of removing irrelevance rather than redundancy of the data.

Information Gain and Canonical Correlation Analysis−Based Gene Selection Algorithm

Yoon et al.[93] proposed an information gain (IG)−related SVM-RFE technique for solving the issue of classification between normal and malignant tissue in breast cancer diagnosis. They have used SVM as the mode of classification with IG as the feature selection method. Sha-Sha et al.[94] suggested a MapReduce algorithm on gene expression datasets for extracting high-scaled gene features for having a greater training accuracy. This method combined a parallel and distributed computing environment with MapReduce algorithm and extreme learning machines. Lei[95] proposed a method that combines IG with genetic algorithm for the purpose of feature selection. The features are chosen based on the IG where the frequency of the item is taken into consideration. To improve the filtering process of the information, a new fitness function was suggested to consider different characteristics of the dataset.

Wei-qiang et al.[96] tried to develop a new modified IG algorithm that overcomes the drawbacks of the traditional IG. The proposed method when applied to different datasets produced a much satisfying results as compared with the existing IG. Sehhati et al.[97] suggested a gene signature model for the gene expression data where a new scoring method was presented based on the mutual information concept. They picked out the genes based on the information content by using the forward and backward gene selection set. Wu et al.[98] proposed a modified feature selection approach based on the IG content where some improvements were made to suit the varied datasets used. The features extracted were tested using the SVM classifier. Shalout et al.[99] proposed a method that combined the PCA method with the IG method for the classification of influenza dataset. It was noted that when the PCA method was alone used for feature selection, the results brought forth were not quite efficient as compared with the combined method. The classification accuracy was slightly compromised, but performance-wise the proposed algorithm was much appreciated. Azhagusundari et al.[100] proposed a method based on the discernibility matrix and IG for the purpose of feature selection in large multidimensional datasets. The outcomes received from the same were quite good as compared with the original IG algorithm.

Correa et al.[101] offered a modification in the canonical correlation analysis (CCA) method for the detection of irregularities in the merger of two or more datasets. The proposed scheme worked better in the feature level and showed drastic changes in the subjects with schizophrenia. Yan et al.[102] proposed a two-dimensional CCA method called S2DCCA method using a low rank approximation method that selects the best features from the data available. The method proved to be effective enough for pulling the best discriminating features. Li et al.[103] proposed a *Gaussian Bayesian network* based on CCA for solving the problem of low efficiency and reliability. The experimental results showed and proved that the CCA method when merged with the Bayesian framework produced a real and actual relationship

between different features even though the conditional variance was large enough. Zhang et al.[104] suggested a sparse representation–based feature selection method based on the CCA technique. This technique was known as group sparse CCA method. It used two sets of variables and aimed at preserving the characteristics of the data.

Gao et al.[105] presented an interesting approach for extracting multiple feature information. For this, they proposed a discriminative multiple CCA technique for deriving discriminative features. Zu et al.[106] used CCA technique for extraction of feature and information in a dataset. The CCA technique was renamed and called as canonical sparse cross-view correlation analysis. They used many kernel functions for the purpose of revealing nonlinear correlation among the sets of features selected. Wang et al.[107] presented the CCA technique for the purpose of information fusion and feature extraction. Their general CCA method was altered and a new approach was suggested that was recognized as canonical principal angle correlation analysis. Here, they ensured that the general process of CCA where the data used to get paired up during the training process is eliminated. Xing et al.[108] thought of presenting CCA in a new alternative mode. There are datasets that are rather high in dimensions. For these datasets, applying CCA directly is relevantly complex. Hence, they proposed a modified CCA technique called complete CCA (C3A) where the single agent matrix calculation was decomposed to two eigen matrix sets. This form of CCA calculation removed the computational burden that usually was a part of occurrence in traditional CCA method. Zhai et al.[109] suggested an instance-specific CCA (ISCCA) technique to remove the traditional CCA's limitation of acting as linear algorithm. The proposed ISCCA technique aimed at approximating the nonlinear data by computing instance-specific data.

Trace Ratio and Signal-to-Noise Ratio–Based Gene Selection Algorithm

Zhang[110] made a study of the trace ratio (TR) concept for linear discriminant analysis (LDA). They attempted to study the TR for uncorrelated constraints. In other words, they made a study of uncorrelated TR for LDA for applications related to undersampled problems. It was remarked that between this proposed approach and variants of LDA, the former performed better as compared with the latter. Zhao et al.[111] proposed a modified TR algorithm for the generalized discriminative learning process for dimensionality reduction.

Here, unlabeled data were used in the training process for conservation of the structural characteristics of the original datasets.

Liu et al.[112] suggested a modified version of the TR criterion for selecting features with small variance factor. They developed a noise-insensitive TR algorithm that worked well when used with the rescaling preprocessing technique. The output derived from the above method outperformed the results produced from any traditional feature selection algorithm. Liu et al.[113] proposed a novel feature selection method called *multiple kernel dimensionality reduction method based on spectral regression and TR criterion*. This method aimed at learning things from the correct kernel for the purpose of transformation into a lower dimensional space. Nie et al.[114] proposed an algorithm where the subset level score was directly optimized and used for finding the global-optimized feature subsets. This algorithm was tested on various algorithms, and in most of the cases, it was proved that the results obtained through this algorithm were significantly better as compared with others. Zhao et al.[115] a dimensionality reduction based TR on the kernel version for handling nonlinear problems. Various types of real and synthetic datasets were used and the same were tested. It was observed that the results produced were quite good as compared with the original TR algorithm.

Huang et al.[116] suggested a TR algorithm based on semisupervised discriminant analysis that used the concept of flexibility. It was called as TR-FSDA. The results obtained from different datasets by using a trace ratio- fast semi-supervised discriminant analysis (TR-FSDA) algorithm were good enough as compared with the other semisupervised dimension reduction methods. Zhao et al.[117] proposed an improved TR-LDA algorithm for dimensionality reduction for dementia datasets. This method was integrated along with the missing value imputation method for analysis of nonlinear datasets in the medical world. Li et al.[118] proposed a TR algorithm with LDA for solving multiclass classification problem for medical data. They used LDA coefficient as conditional probabilities for estimating the posterior probability and calculating the misclassification rate. When compared with other existing feature selection techniques, the proposed method provided a lower total cost and higher accuracy factor.

Gao et al.[119] combined the signal-to-noise ratio (SNR) method in EEG signal analysis for finding out the least amount of interference in the signals. They examined the signals in three types of electrodes and proved that the effect of the three electrodes was better when combined with SNR and the data simulation was also quite easy and effective. Morawski et al.[120] suggested a combined algorithm based on PCA and SNR. The selection of the features was categorically termed as *automatic*, as the selection of the informative features and information was quite fast as compared with other feature selection techniques available. Qian et al.[121] proposed the SNR technique on a number of image scans for covariance spectroscopy. It was observed that the results produced had an interesting relationship with other spectrum signals.

Venet et al.[122] proposed an SNR for microarray data called SNAGE or signal-to-noise applied to gene experiments. The above method is based on the gene-gene correlation and is a fruitful method for measuring the data quality. Hengpraprohm et al.[123] proposed a method for weighing the feature value by SNR score. They compared their outcomes with different feature selections and classifiers, and it was noted that the proposed method achieved good effect in terms of accuracy of the classifier. Goh et al.[124] proposed an SNR method that is hybridized by Pearson correlation coefficient according to the discrimination power toward the classes.

A hybridization technique using independent component analysis and SNR was proposed by Aziz et al.[125] for feature selection or extraction. They expressed that their proposed combined method provided better results with naïve Bayesian classifier as compared with the existing methods. Kourid[126] used parallel *k*-means on MapReduce for clustering features, and then they applied iterative MapReduce that uses SNR as means for ranking clusters. This method provided a better performance in terms of accuracy when applied to large datasets. Maulik et al.[127] proposed a prediction technique by combining fuzzy preference−based rough set method for feature selection with semi-supervised SVMs.

SUMMARY

Here, a detailed discussion was made toward all the relevant fields such as GRN, gene selection, TR, IG, and CCA. Many authors have many significant contributions toward these areas, and a relevant literature has been cited with a brief detail of each. Providing a deeper and significant insight about various gene selection algorithms was greatly needed and desired for. So, various types of technologic advancement in the selection of genes have been stated herewith. In Chapter 3, a new gene selection algorithm is proposed based on the framework of SVM-RFE and Bayesian t-test. This method has been critically assessed, and the classification accuracy for the same has been calculated using different gene values. In all the cases, the results outshined the previous existing techniques.

SVM-BT-RFE: An Improved Gene Selection Framework Using Bayesian t-Test Embedded in Support Vector Machine (Recursive Feature Elimination) Algorithm

SHRUTI MISHRA, PhD • SANDEEP KUMAR SATAPATHY, PhD • DEBAHUTI MISHRA, PhD

INTRODUCTION

The complex patterns of gene expression are engendered in response to specific cellular activities at different levels. One of the most challenging objectives of systems biology is to provide qualitative and quantitative models for reviewing the intricate patterns of gene interaction.[128,129] Among all the models, gene regulatory networks (GRNs)[130] are most crucial. Construction of GRN is the process of identification of genes that interact in a gene-gene interaction network. This helps researchers define the diverse biologic functions and undercurrents of molecular activities taking place in a human body. However, identifying GRNs cannot be accurate because of high density of the network, deficiency of information about biologic organism, and uproar in the expression measurement.[131,132]

Conventionally, gene selection[133,134] is considered as a vital technique with microarray data, and with gene expression data, identification of different cancer classes becomes one of the important tasks for the purpose of better treatment and diagnosis. The DNA microarray technology has provided us several prospects to detect gene expression levels for many thousands of genes simultaneously. The problem is to select a small subsection of genes from a huge pattern of expressions. The challenging task is to choose relevant genes that are highly correlated to classify because of small sample size of the expression data. In other words, gene selection is considered as a primary preprocessing step for cancer classification where redundant and irrelevant genes are eliminated to achieve maximum predictive and decisive accuracy. Gene selection methods are usually categorized into four different approaches: filter,[135] wrapper,[136] embedded,[137] and hybrid.[138] Filter methods[139] evaluate based on the individualities of the data and relation of each gene with the class label. It usually considers statistical properties of the data without any learning model. It is also considered as one of the admired methods, as it is quite simple, efficient, and accurate. The wrapper methods[140] are quite popular in machine learning tasks and applications. This method evaluates the fitness of subset of selected genes iteratively by a specific learning classifier model in the genetic selection process. In the embedded method,[141] using a preliminary gene set, a learning classifier model is skilled to establish a criterion to measure the rank values of genes. As compared with the filter method, these two methods are quite complex and computational pricey, but they usually provide a better accuracy rate (classifier accuracy), as they properly use the features of the classifier for ranking the genes. The hybrid approach[142] takes the benefits of the filter and the wrapper approaches. In the hybrid approach the first subset of genes are chosen based on the filter approach and then the final subset of genes are chosen based on the wrapper approach.

MATERIALS AND METHODS

This section deals with the details of the existing methodologies that are needed and previously exist in required domain. Some of them are stated accordingly.

Bayesian *t*-Test

Ace of the significant statistical difficulties is finding out whether or not a gene is differentially expressed in two different samples. Typically, student *t*-test[143] is used for the purpose, but it possesses some of the major restrictions such as it can only test differences between two groups and it is restricted to a single group or repeated measures designs. To address the deficits of the student *t*-test, many other methods were proposed, of which Bayesian *t*-test[144] was highly debated. A Bayesian framework was used with the *t*-test in the microarray experiments for better accuracy. In simple terms, it is said to provide richer information than a simple *P* value. The Bayesian *t*-test has certain facts that make it more conformist. Like, it is meant to give equal weightage to the null and research hypotheses, it takes sample size into account, etc. It is said to succeed the generalized *t*-test in terms of simplicity and accuracy. It can always be explained in terms of unconditional distributions.

In the framework, the Bayesian formula[145] is used to calculate the posterior probability of any model. The posterior formulation of within-sample variation is given as in Eq. (3.1):

$$\sigma_p^2 = \frac{v_0 \sigma_0^2 + (n-1)\sigma_x^2}{v_0 + n - 2} \tag{3.1}$$

where, σ_x^2 is the real estimate within treatment among replicate variations, n is the number of replicates, and v_0, σ_0^2 is the prior degrees of freedom/variance. In the Bayesian *t*-test, S is given as in Eq. (3.2):

$$S = \frac{\overline{x_1} - \overline{x_2}}{\sigma_p} \tag{3.2}$$

where, $\overline{x_1}$ is the mean of sample 1 and $\overline{x_2}$ is the mean of sample 2.

Support Vector Machine-Recursive Feature Elimination

One of the most critical research problems is selecting genes from thousands of genes and a small number of samples. Support vector machine-recursive feature elimination (SVM-RFE)[76] is a state-of-the-art algorithm that is used for gene selection or gene ranking. Here, SVM's weight vector is used as an idea for producing the feature rank. The position of the features of a classification problem can be provided by the SVM-RFE algorithm as proposed by Guyon et al.[76]

The algorithm is trained by SVM with a linear kernel (depending on the data considered for investigation), and the features are removed recursively using the smallest ranking criterion. As stated earlier, to generate a rank, the weight vector needs to be calculated as given in Eq. (3.3):

$$W = \sum_{i=1}^{n} \beta_i x_i y_i \tag{3.3}$$

where i is the number of genes ranging from 1 to n, β_i is the *Lagrange multiplier* or *support vectors* estimated from the training set, x_i is the gene expression vector for sample i, and y_i is the class label of i ($y_i \in [-1, +1]$). The weight vector W is a combination of linear patterns where most of the weight vectors are found to be zero. The one that is nonzero is the *support vector*, which is also known as *Lagrange multiplier*. The SVM-RFE algorithm is stated in Algorithm 3.1.

Support Vector Machine-*t*-Statistics-Recursive Feature Elimination

SVM-RFE[76] recursively eliminated the genes based on the weight vectors and generated the rank score list. Support vector machine-*t*-statistics-recursive feature elimination (SVM-T-RFE) approach proposed by Li et al.[77] is an enhanced version of the existing SVM-RFE algorithm that incorporates the original SVM-RFE and Welch *t*-test statistics. A two-sample Welch *t*-test with unequal variance is applied along with the weight vectors of the SVM and threshold parameters to bring forth a new modified rank score as given in Eq. (3.4):

$$Rank, r_i = \theta \times w_i + (1 - \theta) \times t_i \tag{3.4}$$

where r_i is the rank of the ith gene, θ is the parameter determining the trade-off between SVM weights and *t*-statistics ranging from 0 to 1, t_i is the Welch *t*-test of the ith gene. Welch *t*-test is defined as in Eq. (3.5):

$$t = \frac{\overline{x_1} - \overline{x_2}}{\sqrt{\dfrac{s_1^2}{n_1} + \dfrac{s_2^2}{n_2}}} \tag{3.5}$$

where n_1 and n_2 are the sizes of sample 1 and sample 2, $\overline{x_1}$ and $\overline{x_2}$ are the means of sample 1 and sample 2, and s_1^2 and s_2^2 are the variances of sample 1 and sample 2. Here, the performance of the gene selection algorithm depends entirely on the θ parameter, as the values are chosen from the set of $\{0 \ldots 1\}$ with a stepwise difference of 0.01. When θ is set to 0, then SVM weight vectors are not considered; when θ is set to 1, then *t*-statistics are not considered. It means in either case, the SVM-T-RFE becomes equal to the SVM-RFE classifier. Apart from this, there is one more rule of selecting this θ value. If from the minimum subsets of genes the number of

---

Input: Initial gene subset, $G = \{1, 2 \ldots n\}$

Output: Rank list according to smallest weight criterion, R.

Step 1: Set R= { }

Step 2: Repeat steps 3-8 until G is not empty

Step 3: Train the SVM using G.

Step 4: Compute the Weight Vector using eq (3.3)

Step 5: Compute the Ranking Criteria, $Rank = W^2$

Step 6: Rank the features as in sorted manner.

$$New_{rank} = sort(Rank)$$

Step 7: Update the Feature Rank list

$$Update \ R = R + G \ (New_{rank})$$

Step 8: Eliminate the feature with smallest rank

$$Update \ G = G - G \ (New_{rank})$$

Step 9: End

ALGORITHM 3.1 SVM-RFE.[76]

genes is the same, then a minimum subset is chosen where θ is small. In any case, the smaller the θ, the more the possibility of selecting *t*-statistics, and hence more differentially expressed genes would be considered. The SVM-T-RFE algorithm is mentioned in Algorithm 3.2.

JUSTIFICATION OF THIS CHAPTER

There is a basic fundamental gene selection technique called SVM-RFE, which usually uses a recursive reduction process for selecting and ranking genes. However, the major issue that lies over this is that the elimination process requires removal of one single gene at a time that is presumably time-consuming. Another extended version of this SVM-RFE technique called SVM-T-RFE was proposed using *t*-test score where the removal of single gene at one iteration was replaced with a fixed set of number of genes based on the total number of genes present in the dataset. However, again this method was heavily dependent on a threshold parameter that lies in the range of 0−1 with a stepwise

---

Input: Initial Gene set, $G= \{1, 2 \ldots n\}$

Parametric value, θ

Output: Ordered gene set based on rank score, r.

Step 1: Set r= { }

Step 2: Compute and normalize *t*-test for genes in G using eq (3.5)

Step 3: Repeat step 4 to 8 until set G is null.

Step 4: Train linear SVM with input as G.

Step 5: Compute the weight vector using eq (3.3)

Step 6: Calculate the ranking score r_i using eq (3.4)

Step 7: Choose the gene with lowest ranking score, $g = min\{r_i\}$

Step 8: Eliminate the minimum ranked gene from the gene set and update the

new gene set, G i.e. $G = G - g$

Update the rank list by adding the eliminated gene from the gene set

$r = r + g$

Step 9: End

ALGORITHM 3.2 SVM-T-RFE.[77]

difference of 0.01. Here, the number of iterations is directly dependent on and proportional to the threshold range, which again makes the whole process tedious and time-consuming. Our proposed method overcomes all these issues, and the details of the same are mentioned in the following sections.

SUPPORT VECTOR MACHINE-BAYESIAN t-TEST-RECURSIVE FEATURE ELIMINATION FOR GENE SELECTION

The statistical Bayesian t-test and SVM-RFE are two foremost techniques that can be used for the genetic selection process. The Bayesian t-test when used alone does not result in an optimal solution, and the same can be stated about the SVM-RFE algorithm. As we know that the P values are engendered from the statistical test with a standard significance level of 5% or 0.05, the differentially expressed genes too can be exemplified from the same. The topmost genes can be used in the ranking criteria along with the two techniques. The ranking criterion is stated as in Eq. (3.6):

$$Rank, R = T_g[\eta \times W_i + (1 - \eta) \times B_{ti}] \quad \textbf{(3.6)}$$

Where T_g is the P value of the topmost genes generated from the Bayesian t-test, η is the parametric accord between SVM weight and Bayesian t-test score, W_i is the SVM weight vector for ith gene, and B_{ti} is the Bayesian

t-test value (P value) for all ith genes. The η range varies between 0 and 1 with an increment value of 0.01 or 0.001 depending on the number of topmost genes selected. The ranking score basically depends on two factors such as the T_g and η. The Bayesian t-test values of the topmost genes when merged with η, W, and B_{ti} result in a minimum number of features needed for maximum classification accuracy.

The ranking score, R, is used to acquire an insight about the importance of each gene for classification. In other words, the ranking score is applied to determine the extent of the importance of a gene i for the purpose of sorting. In the proposed algorithm, the objective is to produce a new rank list by recursively removing the genes having the smallest rank from the gene set list till no genes are present in the gene set.

The original algorithm of Guoyan et al.[76] has been modified, and the proposed algorithm is stated in Algorithm 3.3.

When compared with the previous two versions, i.e., SVM-RFE and SVM-T-RFE, the ranking criteria for support vector machine-Bayesian t-test-recursive feature elimination (SVM-BT-RFE) do vary. For SVM-RFE the ranking criterion is decided by the square of weight vectors (given in Eq. 3.3), whereas for SVM-T-RFE the ranking criterion is derived with the help of two-sample Welch t-test with unequal variance applied with the weight vectors of SVM and some threshold

---

Input: Initial Genesubset, $G = \{1, 2...n\}$

        Ordered gene set based on rank score, $R = \{\}$

        Bayesian t-test values for number of top most genes, T_g

        Parametric value, η

Output: Ordered gene set based on rank score, R.

Step 1: Repeat steps 2-7 until G is not empty

Step 2: Train the SVM using G.

Step 3: Compute the Weight Vector using eq (3.3)

Step 4: Compute the Ranking Criteria using eq (3.6)

Step 5: Rank the features as in sorted manner

        $New_{rank} = sort(Rank)$

Step 6: Update the feature rank list

        Update $R = R + G (New_{rank})$

Step 7: Eliminate the feature with smallest rank

        Update $G = G - G (New_{rank})$

Step 8: End

ALGORITHM 3.3 Proposed SVM-BT-RFE.

parameter (given in Eq. 3.4). The ranking criterion for SVM-BT-RFE is generated by considering the P value of few topmost differentially expressed genes derived from Bayesian t-test along with the P value of all the ith genes and SVM weight vectors for all the ith genes.

EXPERIMENTATION

In this section, we start out with a discussion based on the types of datasets used and the preprocessing of the five datasets to normalize the values followed by the parameters discussion. The schematic model is presented in the following section to state the flow of the algorithm that has been used. For the evaluation of the work, MATLAB® version of R2014a was used with the system requirement of 8 GB RAM.

Datasets Used

Expression profiling of colon cancer or colorectal adenomas and normal mucosas from 32 patients was downloaded from Gene Expression Omnibus.[146] This set consists of 32 adenomas and 32 normal mucosas (64 samples) having 43,237 genes. To illustrate the molecular developments underlying the alteration of normal colonic epithelium, the transcriptomes of 32 prospectively collected adenomas were equated with those of normal mucosas from the same entities. Similarly, the leukemia dataset was collected from Ref. 147 where the dataset is said to consist of 10,056 genes with 48 samples of both ALL (acute lymphocytic leukemia) and AML (acute myeloid leukemia) (24 ALL and 24 AML each). Apart from these two, few more datasets were taken into consideration such as the medulloblastoma dataset[148] having 5893 genes with 34 samples of 25 C and 9 D (medulloblastoma has four molecular subtypes, of which two less well-defined subtypes are group C and group D), lymphoma dataset[149] having 7070 genes with 77 samples of 58 DLBCL (diffuse large B cell lymphoma) and 19 FL (follicular lymphoma) (Affymetrix HuGeneFL array), and the prostate cancer dataset[150] having 12,533 genes with 102 samples of 50 normal and 52 tumor (Affymetrix Human Genome U95Av2 Array platform). These large-scale gene expression datasets were first considered to be statistically evaluated and then were used for the comparison of existing SVM-RFE, SVM-T-RFE, and SVM-BT-RFE algorithms.

Preprocessing of the Data

One of the most essential stages of preprocessing is normalization. This helps to transform the raw data into data appropriate for innumerable application. Datasets were normalized using the *z-score normalization*

or *zero-mean normalization* technique.[151] Here, the values for an attribute are normalized using the mean and the variance. The above technique is stated as in Eq. (3.7):

$$v_i' = \frac{v_i - \overline{R}}{std(R)} \tag{3.7}$$

where v_i' is the *z-score normalized* value of v_i; v_i is the value of the row R of the ith column. The $std(R)$ is the standard deviation given as in Eq. (3.8) and $\overline{R}$ is the mean given as in Eq. (3.9).

$$std(R) = \sqrt{\frac{1}{(n-1)} \sum_{i=1}^{n} (v_i - \overline{R})^2} \tag{3.8}$$

$$\overline{R} = \frac{1}{n} \sum_{i=1}^{n} v_i \tag{3.9}$$

Parameter Discussion

Algorithm 3.3 depicts the proposed SVM-BT-RFE method, where the linear SVM has been trained in each iteration, depending on different sets of G values. The parameters used here are T_g and η. As discussed earlier, the topmost genes are chosen based on the order of descending t-score value or in the order of ascending P value. The top 50, 100, 150, or 500 genes can be selected for the purpose of rank generation. The η value depends drastically on the number of genes selected. The range of η is between 0 and 1 with a difference of either 0.01 or 0.001 or 0.0001 depending on the number of genes that have been selected. The finite set of η should be as $\{0, 0.01…0.99, 1\}$ or $\{0, 0.001, 0.002…1\}$ and likewise. An essential reason for considering the topmost genes is to limit the number of iterations, based on which the ranks can be generated. Instead of considering many iterations, based on the threshold parameter η (i.e., from 0 to 1), the number of iterations gets restricted and the process limits to the number of topmost genes selected. This basically is one of the most important factors considered in our algorithm, which resolves the issue of time consumption that persisted in SVM-RFE and SVM-T-RFE. Now considering the proposed ranking criterion, it can be seen that the T_g is used with parameterized equation, which is a combination of η, w_i and Bt_i. When η is 0, then we only consider the Bt_i values along with the T_g. Similarly, when we consider η to be 1, then we only consider the SVM weight vector values along with the T_g. We get a different ranking criterion for the proposed SVM-BT-RFE when SVM weight vectors or Bayesian t-test values are used alone with the differentially expressed gene sets. The SVM-BT-RFE algorithm also restricts the consumption of time if few numbers of topmost genes are considered.

Implementation and Performance Analysis

The proposed model for the work is described in Fig. 3.1:

The SVM-BT-RFE algorithm (Algorithm 3.3) was realized through a series of steps. To begin with, the datasets using z-score normalization process. Using the normalized datasets, a statistical technique known as Bayesian t-test was used for the determination of gene selection. The Bayesian t-test with a significance level of 5% or 0.05 resulted in a P value, based on which the overexpressed and underexpressed differential genes were mined. The genes having P value less than or equal to 0.05 (5% significance level) were designated as the genes of interest and the remaining were discarded. The first phase of gene selection settles over here.

In the subsequent phase, the normalized dataset was considered to train the linear SVM. The linear SVM yields a *Lagrange multiplier* value β that was further used to calculate the weight vector using Eq. (3.3). These weight vectors were later used for determining the ranking score for the process of recursive elimination of genes. In this approach, before finding the ranking score, the topmost genes were sorted out using the P values (genes having the P values less than or equal to 0.05). In this work, top 50, 100, 150, and 500 genes are used for evaluation.

Beginning with the last and the final process of gene selection, we used Algorithm 3.3. The ranking score was generated using a threshold η ranging from 0 to 1 having a suitable difference of 0.1, 0.01, or 0.001 depending on the number of topmost genes selected. The ranking iteration was dependent on the number of genes selected. Based on the results produced by these rank iterations, 100 genes were eliminated in one pass, as the number of genes was more than 10,000. For genes less than 10,000, 5 genes were removed at a time. This helped to limit the time consumed, which would have been extremely high if one gene was removed at a time. This procedure was iterated until the ordered set G was empty. This process produced a minimum number of genes required for acquiring an accuracy of 100% or an error count of 0.

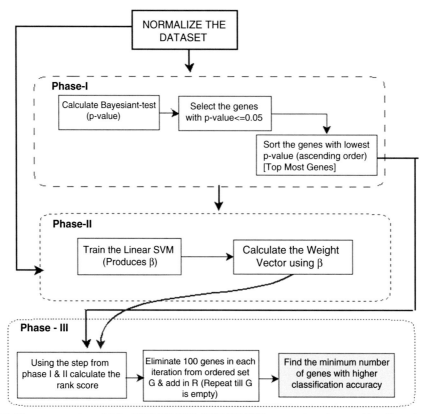

FIG. 3.1 The proposed model for SVM-BT-RFE. *SVM-BT-RFE*, support vector machine-Bayesian t-test-recursive feature elimination.

RESULT ANALYSIS

For the proper discussion of the algorithms, in this work, five different datasets have been used and evaluated from different sources. Table 3.1 presents the entire details and description of the datasets that have been taken for the performance analysis and comparisons.

One of the major issues in the process of gene selection is that, with a huge ordered gene subset, we have to find a smaller subset with higher classification accuracy. Based on this aspect of gene selection, when we compared the original SVM-RFE, SVM-T-RFE, and the proposed SVM-BT-RFE technique for gene selection, it was found that the SVM-BT-RFE technique provided better results with respect to the number of genes selected and accuracy in classification. The performance of all the three algorithms was compared using the five well-known datasets. Tables 3.2–3.6 show the performance of each algorithm with respect to the genes selected and the accuracy obtained using SVM classification.

In Table 3.2, the performance of colon cancer data was found to be better in the SVM-BT-RFE by selecting genes above 700 (97% accuracy). Although for a smaller set of genes the result produced was not satisfying in our algorithm, as compared with the other two the performance of our algorithm was far better. The SVM-RFE shows the least accuracy as compared with the SVM-BT-RFE and SVM-T-RFE, as it recursively removes one gene per iteration. In Table 3.3, the performance of leukemia data was quite appropriate, as maximum accuracy (100%) has been achieved with only six sets of genes using the SVM-BT-RFE. Using the SVM-T-RFE, maximum accuracy was achieved, but a minimum of eight genes were required. Table 3.4 describes the performance measure of medulloblastoma data where with only nine subsets of genes the SVM-BT-RFE and SVM-T-RFE produced maximum achievable accuracy (100%). Table 3.5 presents the analysis of lymphoma data where a minimum of 16 genes were needed

---

TABLE 3.1
Characteristics and Features of the Dataset Used in Experimental Analysis

Data	Amount of Genes	Number of Samples		Training Data	Testing Data	References
		Class 1	Class 2			
Colon	43,236	32 (N)	32 (T)	45	19	146
Leukemia	10,056	24 (ALL)	24 (AML)	34	14	147
Medulloblastoma	5893	25 (C)	9 (D)	24	10	148
Lymphoma	7070	58 (D)	19 (FL)	54	23	149
Prostate cancer	12,533	50 (N)	52 (T)	44	18	150

ALL, acute lymphocytic leukemia; *AML*, acute myeloid leukemia; *FL*, follicular lymphoma.

TABLE 3.2
Performance Analysis of SVM-BT-RFE As Compared With SVM-RFE and SVM-T-RFE for Colon Cancer Dataset

Amount of Genes	Average Classification Accuracy (in %)								
	SVM-BT-RFE			SVM-T-RFE			SVM-RFE		
	Best	Mean	Worst	Best	Mean	Worst	Best	Mean	Worst
100	68	68	68	55	55	55	50	50	50
300	89.3	87	85	65.01	65	65	53	53	53
500	93	90.89	90	70.23	68.3	67.23	63	62	61.58
700	97	93.45	90.28	74.68	72.99	72.32	72.78	71.68	71
900	99.5	95	93	84.87	80.96	79.05	78	77.03	77.67

SVM-RFE, support vector machine-recursive feature elimination; *SVM-BT-RFE*, support vector machine-Bayesian *t*-test-recursive feature elimination; *SVM-T-RFE*, support vector machine-*t*-statistics-recursive feature elimination.

TABLE 3.3
Performance Analysis of SVM-BT-RFE As Compared With SVM-RFE and SVM-T-RFE for Leukemia Dataset

	Average Classification Accuracy (in %)								
	SVM-BT-RFE			SVM-T-RFE			SVM-RFE		
Amount of Genes	Best	Mean	Worst	Best	Mean	Worst	Best	Mean	Worst
6	100	94.87	87.5	98	92	85	65	58	50
7	100	97	90	99.53	95.10	89.23	68.21	60.01	50.71
8	100	98.01	91.26	100	97	93	75.31	64.2	52.9
9	100	99	93	100	98.02	94.25	80.2	68	54.01

SVM-RFE, support vector machine-recursive feature elimination; *SVM-BT-RFE*, support vector machine-Bayesian *t*-test-recursive feature elimination; *SVM-T-RFE*, support vector machine-*t*-statistics-recursive feature elimination.

TABLE 3.4
Performance Analysis of SVM-BT-RFE As Compared With SVM-RFE and SVM-T-RFE for Medulloblastoma Dataset

	Average Classification Accuracy (in %)								
	SVM-BT-RFE			SVM-T-RFE			SVM-RFE		
Amount of Genes	Best	Mean	Worst	Best	Mean	Worst	Best	Mean	Worst
9	100	97.62	95.23	100	98	96.01	94	89.45	85
12	100	98.1	96.23	100	98.7	97.56	98	97	94
16	100	98.7	97.45	100	99.4	98.89	100	98.3	96.78
18	100	99	98	100	99.45	99	100	98.7	97.54

SVM-RFE, support vector machine-recursive feature elimination; *SVM-BT-RFE*, support vector machine-Bayesian *t*-test-recursive feature elimination; *SVM-T-RFE*, support vector machine-*t*-statistics-recursive feature elimination.

TABLE 3.5
Performance Analysis of SVM-BT-RFE As Compared With SVM-RFE and SVM-T-RFE for Lymphoma Dataset

	Average Classification Accuracy (in %)								
	SVM-BT-RFE			SVM-T-RFE			SVM-RFE		
Amount of Genes	Best	Mean	Worst	Best	Mean	Worst	Best	Mean	Worst
10	98.01	85.12	72.24	97	83.33	69.86	86	76.78	68
13	99.3	89.65	80	98.5	87.05	77	90.24	81.71	75
16	100	94.5	89.04	99.89	93.38	86.87	98	89	81.36
19	100	95.52	91.78	100	95.16	90.33	100	96.01	92.21

SVM-RFE, support vector machine-recursive feature elimination; *SVM-BT-RFE*, support vector machine-Bayesian *t*-test-recursive feature elimination; *SVM-T-RFE*, support vector machine-*t*-statistics-recursive feature elimination.

TABLE 3.6
Performance Analysis of SVM-BT-RFE As Compared With SVM-RFE and SVM-T-RFE for Prostate Cancer Dataset

	Average Classification Accuracy (in %)								
	SVM-BT-RFE			SVM-T-RFE			SVM-RFE		
Amount of Genes	Best	Mean	Worst	Best	Mean	Worst	Best	Mean	Worst
12	**98.11**	**94.33**	**89**	95	93.7	87.45	84	78..2	72.48
15	**98.79**	**96**	**92.14**	96.23	94.6	89.33	93.78	87.39	81.01
18	**99.34**	**97.49**	**95**	97.51	96.6	93.25	95.01	92	86
20	**99.64**	**98.22**	**96.42**	98.4	97	94	98.10	96	91.57

SVM-RFE, support vector machine-recursive feature elimination; *SVM-BT-RFE*, support vector machine-Bayesian *t*-test-recursive feature elimination; *SVM-T-RFE*, support vector machine-*t*-statistics-recursive feature elimination.

out of 7070 genes for attaining 100% accuracy. Table 3.6 presents the prostate cancer dataset for which the SVM-BT-RFE provided highest accuracy with 20 genes, but still a minimum of 12 genes did provide a good accurate value. Although for this dataset the SVM-T-RFE and SVM-RFE too provided quite a good figured accuracy as compared with our proposed algorithm, but with a slight good margin, our algorithm was shown to be more beneficial. From the above tables, we can conclude that the SVM-BT-RFE is providing better results as compared with the other two algorithms.

For performing a detailed comparison of all the three algorithms, we can present different classification error rates that we obtained after about five runs or iterations of the algorithm. Instead of completing all the iterations, we chalked out a small run with a minimal gene subset. Tables 3.7–3.11 present the comparison analysis performed with proper classification error rates obtained with only five iterations or runs.

Instead of executing the algorithms fully, we partially executed all the three algorithms for a maximum of five iterations. It was found that the SVM-BT-RFE algorithm still outperforms the other two algorithms in terms of classification error rates. Table 3.7 presents the classification error rate for the colon cancer dataset from 43,236 genes. As the gene size is large, at one iteration, we removed a maximum of 100 genes. The classification error rate for the above mentioned dataset in SVM-BT-RFE was found to be higher for 100 genes, but it gradually diminished as the size of the genes started increasing. The SVM-T-RFE and SVM-RFE too showed an increased percent of error rates for the abovementioned dataset.

TABLE 3.7
Average Classification Error (in %) Over Five Runs for the Three Gene Selection Methods for Colon Cancer Dataset

	Amount of Genes Selected			
Methods	100	200	300	400
SVM-BT-RFE	**28**	**11.24**	**10.41**	**9.02**
SVM-T-RFE	45.01	37	36.12	34.33
SVM-RFE	53.57	52.5	51.18	50.07

SVM-RFE, support vector machine-recursive feature elimination; *SVM-BT-RFE*, support vector machine-Bayesian *t*-test-recursive feature elimination; *SVM-T-RFE*, support vector machine-*t*-statistics-recursive feature elimination.

TABLE 3.8
Average Classification Error (in %) Over Five Runs for the Three Gene Selection Methods for Leukemia Dataset

	Amount of Genes Selected			
Methods	5	10	15	20
SVM-BT-RFE	**10.02**	**5.14**	**2.41**	**1.56**
SVM-T-RFE	15.25	7.89	3.34	2.05
SVM-RFE	30	17.46	9.78	5

SVM-RFE, support vector machine-recursive feature elimination; *SVM-BT-RFE*, support vector machine-Bayesian *t*-test-recursive feature elimination; *SVM-T-RFE*, support vector machine-*t*-statistics-recursive feature elimination.

TABLE 3.9
Average Classification Error (in %) Over Five Runs for the Three Gene Selection Methods for Medulloblastoma Dataset

Methods	Amount of Genes Selected			
	5	10	15	20
SVM-BT-RFE	**17.58**	**12.41**	**9.89**	**9**
SVM-T-RFE	22	19.87	17.98	15
SVM-RFE	34	32.14	31.05	29.87

SVM-RFE, support vector machine-recursive feature elimination; *SVM-BT-RFE*, support vector machine-Bayesian *t*-test-recursive feature elimination; *SVM-T-RFE*, support vector machine-*t*-statistics-recursive feature elimination.

TABLE 3.11
Average Classification Error (in %) Over Five Runs for the Three Gene Selection Methods for Prostate Cancer Dataset

Methods	Amount of Genes Selected			
	10	15	20	25
SVM-BT-RFE	**30.01**	**27.33**	**24.45**	**21.74**
SVM-T-RFE	33	31.5	29.56	25.86
SVM-RFE	36.10	33.56	31.07	28.30

SVM-RFE, support vector machine-recursive feature elimination; *SVM-BT-RFE*, support vector machine-Bayesian *t*-test-recursive feature elimination; *SVM-T-RFE*, support vector machine-*t*-statistics-recursive feature elimination.

TABLE 3.10
Average Classification Error (in %) Over Five Runs for the Three Gene Selection Methods for Lymphoma Dataset

Methods	Amount of Genes Selected			
	10	15	20	25
SVM-BT-RFE	**29.33**	**27.74**	**22.85**	**18.75**
SVM-T-RFE	33.86	31.87	29.87	26.14
SVM-RFE	36.14	34.20	31.20	28.10

SVM-RFE, support vector machine-recursive feature elimination; *SVM-BT-RFE*, support vector machine-Bayesian *t*-test-recursive feature elimination; *SVM-T-RFE*, support vector machine-*t*-statistics-recursive feature elimination.

Table 3.8 shows the error rate with a leukemia dataset. We took certain number of genes in the dataset and found that the corresponding error rate was related. The SVM-RFE showed the maximum classification error rate in 5, 10, 15, and 20 genes. Table 3.9 presents the medulloblastoma dataset where the SVM-BT-RFE gave a lesser error rate as compared with the SVM-T-RFE and SVM-RFE for 5, 10, 15 and 20 genes. Tables 3.10 and 3.11 depict the lymphoma and prostate cancer datasets. Here too, we can find the SVM-RFE providing maximum classification error rate as compared with the other two techniques. In an overall representation and conclusion drawn from the above tables, it was found that although the SVM-T-RFE provided better results as compared with SVM-RFE, still the SVM-BT-RFE supplied the best results regarding the error rate, accuracy, and selection of number of smaller gene sets as compared with the other two techniques. We can also increase the number of iterations, but the result would be more or less same as stated from Tables 3.7—3.11.

Figs. 3.2—3.6 show the average classification error rate (%) that we obtained in five runs and the respective comparison results for colon cancer, leukemia, medulloblastoma, lymphoma, and prostate cancer datasets for the proposed SVM-BT-RFE, SVM-T-RFE, and SVM-RFE. It has been observed that the proposed algorithm outperformed the remaining two algorithms.

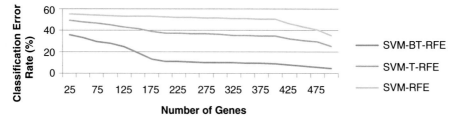

FIG. 3.2 Classification error rates (over five runs) for SVM-BT-RFE, SVM-T-RFE, and SVM-RFE with respect to the number of genes selected for colon cancer dataset. *SVM-RFE*, support vector machine-recursive feature elimination; *SVM-BT-RFE*, support vector machine-Bayesian *t*-test-recursive feature elimination; *SVM-T-RFE*, support vector machine-*t*-statistics-recursive feature elimination.

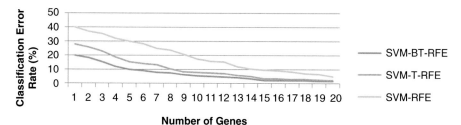

FIG. 3.3 Classification error rates (over five runs) for SVM-BT-RFE, SVM-T-RFE, and SVM-RFE with respect to the number of genes selected for leukemia dataset. *SVM-RFE*, support vector machine-recursive feature elimination; *SVM-BT-RFE*, support vector machine-Bayesian *t*-test-recursive feature elimination; *SVM-T-RFE*, support vector machine-*t*-statistics-recursive feature elimination.

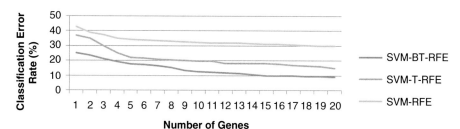

FIG. 3.4 Classification error rates (over five runs) for SVM-BT-RFE, SVM-T-RFE, and SVM-RFE with respect to the number of genes selected for medulloblastoma dataset. *SVM-RFE*, support vector machine-recursive feature elimination; *SVM-BT-RFE*, support vector machine-Bayesian *t*-test-recursive feature elimination; *SVM-T-RFE*, support vector machine-*t*-statistics-recursive feature elimination.

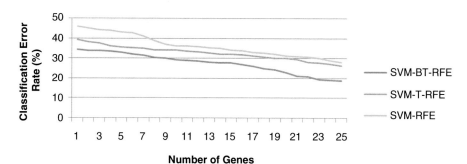

FIG. 3.5 Classification error rates (over five runs) for SVM-BT-RFE, SVM-T-RFE, and SVM-RFE with respect to the number of genes selected for lymphoma dataset. *SVM-RFE*, support vector machine-recursive feature elimination; *SVM-BT-RFE*, support vector machine-Bayesian *t*-test-recursive feature elimination; *SVM-T-RFE*, support vector machine-*t*-statistics-recursive feature elimination.

DISCUSSION

This work can be summarized as follows:

- The original SVM-RFE algorithm used for gene selection aims at eliminating genes recursively. It basically eliminates one gene at a time. Although the algorithm is a state-of-the-art technique, the limitations (consumption of the high amount of training time, elimination of one gene at a time, and

overfitting problem) of the algorithm make it extensively discouraging to use.

- The extended version of the SVM-RFE algorithm called SVM-T-RFE that is a conjunction of SVM-RFE and Welch *t*-test was highly recognized, as it is aimed at training the algorithm in a much faster manner by eliminating many a genes at a time. Based on the size of the dataset, the genes were

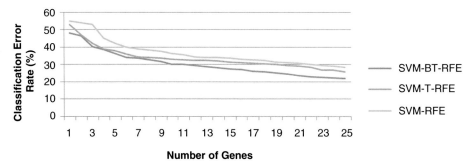

FIG. 3.6 Classification error rates (over five runs) for SVM-BT-RFE, SVM-T-RFE, and SVM-RFE with respect to the number of genes selected for prostate cancer dataset. *SVM-RFE*, support vector machine-recursive feature elimination; *SVM-BT-RFE*, support vector machine-Bayesian *t*-test-recursive feature elimination; *SVM-T-RFE*, support vector machine-*t*-statistics-recursive feature elimination.

recursively removed, making the algorithm faster enough to work with.

- Both the above said algorithm work with the weight vectors to get the rank score. Based on the rank score, the gene subsets were generated that predicted maximum classification accuracy.
- This chapter proposed a further extended algorithm called SVM-BT-RFE where we have considered the merits of SVM-RFE and SVM-T-RFE. The algorithm aimed at generating a minimal subset of genes by predicting the rank score using the recursive elimination technique.
- The algorithm considered the findings of the statistical Bayesian *t*-test and generalized *t*-test and merged it with the weight vector to produce a new ranking score.
- The generalized *t*-test was used as a filter for extracting the topmost genes (most significant genes whose P values were less than the 5% significance level), which were used along with the SVM weights and Bayesian *t*-test P values to develop a faster algorithm.
- The algorithm, although seemed to be a little time-consuming, is quite fast enough, as the number of iterations is scaled down to the number of topmost genes selected and the threshold range of $0-1$.
- Comparing the three algorithms with five datasets, it was found that the SVM-BT-RFE is quite powerful enough in selecting the minimum subset of genes that attain maximum classification accuracy.
- Proof can be easily realized over the graph shown in Figs. 3.2–3.6, where the classification error rates (over five runs) for five datasets with respect to all the three methods have been shown.

SUMMARY

Statistical tests such as Bayesian *t*-test, *t*-test, Welch *t*-test, and many more are quite prominent techniques for finding differentially expressed genes in microarray dataset. The existing SVM-RFE algorithm creates a gene rank list by training a linear SVM and eradicating the gene with lowest ranking criterion in a classification task. Our proposed SVM-BT-RFE technique is likewise intended to generate a rank list of the top-listed genes with the additional straining process (as shown in phase I of Fig. 3.1) using the statistical Bayesian *t*-test. The performance of our proposed algorithm was found to be better than the existing SVM-RFE algorithm, as it produced a minimum number of genes with better prediction accuracy. Although the result generated by the algorithm was far better than the existing technique, the additional filtering process does increase the complexity of the algorithm. The proposed SVM-BT-RFE can be used further in constructing the GRN. Genes selected from the proposed method can be used for network construction which will help us to detect the focused genes that lead to a particular type of disease. In the next chapter, a new gene selection algorithm has been proposed using trace ratio algorithm, which promises to provide a far better result as compared with the SVM-BT-RFE. It also aims at resolving certain issues that occur in the SVM-BT-RFE such as time consumption and dependency factor.

CHAPTER 4

Enhanced Gene Ranking Approaches Using Modified Trace Ratio Algorithm for Gene Expression Data

SHRUTI MISHRA, PhD • SANDEEP KUMAR SATAPATHY, PhD • DEBAHUTI MISHRA, PhD

INTRODUCTION

Genes, as good as their yields (proteins), are the essential construct blocks of animation that do not function autonomously. Rather for a cell to function appropriately, they act together and form an intricate network.[152] Gene networks signify the relationship between sets of genes that harmonize to achieve different tasks. For the understanding of the core biologic process and its molecular system, the gene regulatory network (GRN)[130] plays a crucial part. The ultimate objective of genomics is to understand the causes behind the characteristics of the genetic organisms. However, modeling of these networks is a significant challenge that needs to be addressed.

Apart from this, understanding the construction and functionalities of GRNs is a basic problem in biology. With the accessibility of gene expression data and whole genome sequences, several computational approaches have been developed to set apart their regulatory network by enabling the recognition of the regulatory state component.[153] In other words, with the presence of the gene expression information, identification of the interactions between the genes is quite fast enough using the gene network environment. In the current era, formation of precise GRN models[19] is reaching a major percentage of importance in biomedical research. The gene expression of the microarray data monitors the behavior of thousands of genes simultaneously that provides an enormous chance to look into large-scale regulatory networks. Lastly, an absolute GRN model allows us to incorporate experimental facts about the elements and interactions of the factors, which leads to knowing the final state or the dynamical behavior of the network. In fact it provides both quantitative and qualitative scopes for modeling and simulating the networks that

have the potential to reduce the effect of the topological effect of interactions with other living organisms.

To construct a GRN model, gene selection[128,129] acts as a major criterion. Gene selection from microarray data, which is high-dimensional data, is statistically a difficult problem. It is basically important to select some of the informative genes that carry biologic information in the living organisms. Hence, it is considered as one of the most important data analysis steps. Generally, the number of samples is less as compared with thousands of genes whose expression levels are measured. Hence, it is important to restrain down to few disease-related genes from thousands of microarray genes by the operation of *selection* or *ranking*. Apart from this, selection of optimal number of genes is highly needed for the purpose of classification, as we would be able to select the best 50 ranked genes or 100 ranked genes so as to prove that, with the smaller set of gene sets, the classification accuracy rate is superior.

There are many gene selection or feature selection methods[133,134] that deal with the problem of curse of dimensionality in microarray data. Apart from this, it also helps to reduce the time and memory complexities, which always create issues. Generally, gene selection or feature selection methods are split into two categories: classifier independent and classifier dependent. Filter methods[135] are believed to be classifier dependent, as the choice of the future is based on some heuristic criterion and score, whereas wrapper and embedded methods are thought to be a part of the classifier-dependent method. Wrapper method[136] assesses a subset of variables according to their efficacy to a given predictor, whereas in embedded methods, a variable selection is performed as a part of the learning practice and is usually precise to a given learning machine.

In this chapter, we have proposed two methods in which the trace ratio (TR) algorithm has been explored properly. In our first method, we have not altered any criteria of TR algorithm. Rather, it has been improvised and the dataset was structured on the basis of information gain (IG) values. In our second method, we have modified the existing and original TR algorithm by changing the scoring criteria, which is one of the fundamental steps in the TR algorithm. Instead of using the Fisher score, the canonical correlation analysis (CCA) score is used to calculate the weight matrices within class and between class. Both the proposed methods have been examined and evaluated on the basis of five datasets, i.e., colon cancer,[154] leukemia,[147] medulloblastoma,[148] lymphoma,[149] and prostate cancer.[150] The nature of the dataset is quite large enough in terms of the number of genes, but the dataset has a small sample size. It has been found that the IG with the original TR algorithm and the modified TR algorithm provided fabulous results as compared with the unmodified TR algorithm. Moreover, the canonical correlation score being a statistical technique aims at providing a better rank list when merged with the TR algorithm as compared with the existing Fisher score. It is also relevant, as it is expected to provide a far better classification accuracy rate when compared with the original TR algorithm.

PRELIMINARIES

This section depicts the materials and methods and some of the preliminary facts that have been used for this work such as the methods and the algorithms used such as IG, TR algorithm, and CCA. Some of the existing performance metrics have also been defined such as Kuncheva stability index (KSI), balanced classification rate (BCR), and balanced error rate (BER).

Information Gain

Information gain[155] is a synonym for Kullback-Leibler deviation. On the other hand, in the context of decision trees, the phrase is sometimes used synonymously with mutual information, which is the prospected value of the Kullback-Leibler divergence of a conditional probability distribution. Further IG ratio can be elaborated as the ratio of IG to the inherent information. It is used to diminish a bias toward multivalued attributes by enchanting the number and size of branches into account when choosing an attribute.[156] One of the most vital characters of this is to bias the decision tree against considering attributes with large number of distinct values. That is, it helps in deciding which attributes are the most relevant. IG being an important concept in information theory is applied in the field of machine

learning. In a classification system, for microarray data the IG[157] is designed for each gene. A gene of an arithmetical amount of information provided in the classification system to decide the classification system for the gene of importance. This method can quickly rule out a large number of noncritical noise and inappropriate genes and process the search area of the most favorable subset of genes. Entropy is the measure that is used to reckon the information and compute the degree of vagueness of a random variable. Let node N represents or holds tuples of partition D. The expected information needed to classify a tuple D is given by Eq. (4.1):

$$Info(D) = -\sum_{i=1}^{m} p_i \, log_2(p_i) \qquad (4.1)$$

where $Info(D)$ is the entropy of D and p_i is the probability that an arbitrary tuple in D belongs to class C_i. Suppose the tuples D are partitioned on some attribute A having v distinct values $\{a_1, a_2,...a_n\}$. If A is discrete valued, then it can correspond directly to the v outcomes of a test on A. Attribute A can be used to split D into v partitions or subsets $\{D_1, D_2,...D_n\}$, where D_j contains those tuples in D that have outcome a_j of A. This amount can be measured as shown in Eq. (4.2):

$$Info_A(D) = \sum_{j=1}^{v} \frac{|D_j|}{|D|} \times Info(D_j) \qquad (4.2)$$

Here, the term $\frac{|D_j|}{|D|}$ acts as the weight of the jth partition. $Info_A(D)$ is the expected information required to classify a tuple from D based on the partitioning by A. The IG is the difference between the original information required that is based on the proportion of classes and the new requirement obtained after partitioning A. This is shown in Eq. (4.3):

$$Gain\,(A) = Info(D) - Info_A(D) \qquad (4.3)$$

The larger the divergence, the stronger the correlation. As a result, the differential entropy–defined IG (Algorithm 4.1) represents the quantity of information obtained after the exclusion of uncertainty. Evidently, the larger IG value a feature has, the larger contribution it makes, which is more vital for the classification. Hence, when choosing genes, the one with great IG is selected to represent the original high-dimensional gene first and use them as a base for supplementary gene selection.

Trace Ratio

Feature reduction is a main issue in many machine learning and pattern recognition applications, and the TR problem is an optimization setback concerned in

Input: Original dataset, *D*

Output: Reordered gene sets as per the information gain values

obtained for each attributes in *D*.

Step 1: Find the probability of each category of known samples.

Step 2: Compute the entropy of the classification system (using eq (4.1)).

$$Info(D) = -\sum_{i=1}^{m} p_i \, log_2(p_i)$$

Step 3: Compute the probability and computational probability of all values

for each gene.

Step 4: Calculate the conditional entropy or expected information required

for classifying a tuple from *D* (using eq (4.2)).

$$Info_A(D) = \sum_{j=1}^{v} \frac{|D_j|}{|D|} \times Info(D_j)$$

Step 5: Compute the information gain for all genes (using eq (4.3)).

$$Gain(A) = Info(D) - Info_A(D)$$

Step 6: Sort the results obtained in step 5 based on the descending

order of the gain obtained.

ALGORITHM 4.1 Information gain.[155]

many dimensionality reduction algorithms. Traditionally, the solution is approximated via generalized eigen value decomposition because of the intricacy of the original problem. Fisher and Laplacian score[158,159] are the two famous gene selection algorithms that belong to the graph-based gene selection environment. TR[160] is one of them, i.e., it is a graph-based gene or feature selection algorithm that uses the two scores (Fisher and Laplacian score) as the evaluation criteria measure.

Let us consider two undirected graphs G_w and G_b for within-class and between-class relations that are constructed using Fisher score, where the equivalent adjacency matrices are W_w and W_b. For a dataset X, where both the instances x_i and x_j belong to the same class, the within-class relationship will be higher. So, the feature subset selection should minimize (Eq. 4.4),

$$\sum_{ij} \|l_i - l_j\|^2 (M_w)_{ij} \tag{4.4}$$

for the same class, otherwise maximize. Between-class relationship for both x_i and x_j will be higher when

they belong to different classes. So, the selected gene or feature subset should maximize (Eq. 4.5),

$$\sum_{ij} \|l_i - l_j\|^2 (M_b)_{ij} \tag{4.5}$$

for the different classes, otherwise minimize. Here, l_i is the instance of class for x_i. To find the weight matrices M_w and M_b, Fisher score or Laplacian score is used based on whether it is supervised or unsupervised feature selection. The weight matrices for Fisher score can be classified as given in Eqs. (4.6) and (4.7):

$$(M_w)_{ij} = \begin{cases} \dfrac{1}{num_{l_i}}, & \text{if } l_i = l_j \\ 0, & \text{if } l_i \neq l_j \end{cases} \tag{4.6}$$

$$(M_b)_{ij} = \begin{cases} \dfrac{1}{num} - \dfrac{1}{num_{l_i}}, & \text{if } l_i = l_j \\ \dfrac{1}{num}, & \text{if } l_i \neq l_j \end{cases} \tag{4.7}$$

where l_i denotes the class label of the ith instance of x_i and num_{l_i} denotes the number of data or records belonging to class l_i. The adjacency matrix using Laplacian score can be computed as shown in Eqs. (4.8) and (4.9):

$$(M_w)_{ij} = \begin{cases} e^{\frac{\|x_i - x_j\|^2}{t}}, & \text{if } x_i \text{ and } x_j \text{ are neighbours} \\ 0, & \text{otherwise} \end{cases} \quad (4.8)$$

$$(M_b)_{ij} = \left\{ \frac{1}{1^T DM_w 1} DM_w 1 1^T DM_w \right. \quad (4.9)$$

where Eq. (4.8) denotes the radial distance and t denotes any constant. To unite both the objectives in a single function, ratio of the two is considered and maximized. The ratio is given by Eqs. (4.10) and (4.11):

$$\varphi(S_p) = \frac{\sum_{ij} \|l_i - l_j\|^2 (M_b)_{ij}}{\sum_{ij} \|l_i - l_j\|^2 (M_w)_{ij}} \quad (4.10)$$

$$\varphi(S_p) = \frac{tr\left(S_p^T X L M_b X^T S_p\right)}{tr\left(S_p^T X L M_w X^T S_p\right)} \quad (4.11)$$

where $S_p = [s_{i_1}, s_{i_2}, \ldots s_{i_k}]$ denotes the selection matrix, where $i_1, i_2, \ldots, i_k$ are the first k elements of the transformation $[1, 2, \ldots, n]$, which is gene or feature number. s_{ir} denotes a column matrix with all zeros excluding 1 in the rth position and tr is the TR of the matrix. Let LM_w and LM_b are Laplacian matrices of the form given in Eqs. (4.12) and (4.13):

$$LM_w = DM_w - M_w \quad (4.12)$$
$$LM_b = DM_b - M_b \quad (4.13)$$

where DM_w and DM_b are diagonal matrices given in Eqs. (4.14) and (4.15).

$$(DM_w)_{ii} = \sum_{ij} (M_w)_{ij} \quad (4.14)$$

$$(DM_b)_{ii} = \sum_{ij} (M_b)_{ij} \quad (4.15)$$

Let $Y = XLM_b X^T$ and $Z = XLM_w X^T$. The score of the feature or gene set is calculated as per the TR criteria for a particular selection matrix S_p, which is given as in Eq. (4.16),

$$\beta = \varphi(S_p) = \frac{tr\left(S_p^T Y S_p\right)}{tr\left(S_p^T Z S_p\right)} \quad (4.16)$$

Score of each gene or feature f_i is computed using Eq. (4.17),

$$F(f_i) = m_i^T (Y - \beta Z) m_i \quad (4.17)$$

where m_i is the column vector with all zeros except 1 and the ith position and F is the selected feature or gene set. The algorithm of the TR is stated in Algorithm 4.2.

Canonical Correlation Analysis

CCA[161] is a well-liked statistical method that has been broadly used in information union to confine the correlation between two variables. CCA is an algorithm that is essentially used to dig up the discriminate feature, or genes, and lessen the superfluous information for gene selection. It is also a well-known multivariate analysis method for quantifying the correlation between two sets of multidimensional variables.[162] One of the main intents of CCA is to find and enumerate the correlation between two sets of multidimensional variables. It uses two views of the same pattern and projects them onto a lower-dimensional space in which they are maximally correlated. The traditional CCA algorithm requires to calculate both the inverse and eigen decomposition of a $D \times D$ matrix.[163]

Let us consider two sets of variables X and Y, which contains r variables in set X and q variables in set Y.

$$X = \begin{pmatrix} X_1 \\ X_2 \\ \vdots \\ X_r \end{pmatrix} \text{ and } Y = \begin{pmatrix} Y_1 \\ Y_2 \\ \vdots \\ Y_q \end{pmatrix}$$

We pick X and Y based on the number of variables that subsist in each set so that $r \leq q$. A set of linear combinations called U and V is defined where U corresponds to X and V corresponds to Y of linear combinations. Each member of U will be paired with a member of V. This leads to the sets of section as given below:

$$U_1 = a_{11} X_1 + \cdots + a_{1r} X_r$$
$$\vdots$$
$$U_r = a_{r1} X_1 + \cdots + a_{rr} X_r$$
$$V_1 = b_{11} Y_1 + \cdots + b_{1q} Y_q$$
$$\vdots$$
$$V_r = b_{r1} Y_1 + \cdots + b_{rq} Y_q$$

Hence, (U_i, V_i) is defined as the ith canonical variate pair. The variance of U_i variables can be computed using Eq. (4.18):

$$var(U_i) = \sum_{k=1}^{r} \sum_{l=1}^{q} a_{ik} a_{il} cov(X_k, X_l) \quad (4.18)$$

Similarly, the variance of V_j is computed using Eq. (4.19):

$$var(V_j) = \sum_{k=1}^{r} \sum_{l=1}^{q} b_{jk} b_{jl} cov(Y_k, Y_l) \quad (4.19)$$

Step 1: Calculate adjacency matrices for within the class (M_w) and between the classes (M_b) using Fisher score as follows (eq (4.6) and eq (4.7)):

$$M_w = \frac{1}{num_{l_i}}, if\, l_i = l_j \wedge 0,\ if\, l_i! = l_j$$

$$M_b = \frac{1}{num} - \frac{1}{num_{l_i}}, if\, l_i = l_j \wedge \frac{1}{num}, if\, l_i! = l_j$$

Step 2: Calculate the diagonal matrices (DM_w and DM_b) for the above adjacency matrices as given below (as in eq (4.14) and eq (4.15)):

$$(DM_w)_{ii} = \sum_{ij}(M_w)_{ij}$$

$$(DM_b)_{ii} = \sum_{ij}(M_b)_{ij}$$

Step 3: Calculate Laplacian matrices (LM_w and LM_b) using the eq (4.12) and eq (4.13).

$$LM_w = DM_w - M_w$$

$$LM_b = DM_b - M_b$$

Step 4: Construct a matrix of k features by initially selecting randomly k features from original dataset (say R_k).

Step 5: Declare an empty matrix (say N_k)to store top k features after finding scores of each feature

Step 6: Repeat steps 6 to 10 until $R_k! = N_k$

Step 7: Calculate $Y = XLM_bX^T$ and $Z = XLM_wX^T$

Step 8: Calculate Trace Ratios as $TR_y = TR(R_k^TYR_k)$ and $TR_z = TR(R_k^TZR_k)$

Step 9: Calculate $\beta = \frac{TR_y}{TR_z}$

Step 10: Calculate score of each feature as $F(f_i) = m_i^T(Y - \beta Z)m_i$

Step 11: Select new top k features based on the score and store in N_k

Step 12: Store final k features R_k for further processing

Step 13: Stop

ALGORITHM 4.2 Trace ratio.[160]

Now, covariance between U_i and V_j can be computed as shown in Eq. (4.20):

$$cov(U_i, V_j) = \sum_{k=1}^{r} \sum_{l=1}^{q} a_{ik} b_{jl} cov(X_k, Y_l) \qquad (4.20)$$

The canonical correlation between U_i and V_j can be calculated using Eq. (4.21):

$$\rho_i = \frac{cov(U_i, V_j)}{\sqrt{var(U_i) var(V_i)}} \qquad (4.21)$$

Performance Metrics

Stability of the selected features is a significant aspect when the task is knowledge discovery, not simply returning an accurate classifier. For the validation and assessment of the proposed methods, three different forms of metrics have been applied. Although there are several validation indexes that are available, but for our domain, KSI,[164] BCR,[165] and BER[166] have been used. The detailed explanation of the three metrics is given below:

1. Kuncheva Stability Index[164]

   Let the number of features be in two subsets A and B. KSI is a stability measure that assumes that A and B have the same size (cardinality), i.e., $|A| = |B| = k$ where k denotes the number of features in A or B. In other words, for two subsets, A, $B \subset X$ such that $|A| = |B| = k$ and $r = |A \cap B|$, where $0 < k < |X| = n$ is (as shown in Eq. 4.22),

   $$C(A, B) = \frac{observed\ r - Expected\ r}{Max\ r - Expected\ r} = \frac{rn - k^2}{k(n - k)} \qquad (4.22)$$

   KSI is the average of pairwise consistency. A value 0 indicates the highest possible instability, whereas value 1 indicates the highest possible stability, i.e., all feature subsets have the same cardinal value and all subsets are identical.

2. Balanced classification rate[165]

   BCR is the mean of sensitivity and specificity that introduces a balance amid the classification of two classes (as shown in Eq. 4.23).

   $$BCR = \frac{1}{2}(Sensitivity + Specificity) = \frac{1}{2}\left(\frac{TP}{TP + FN} + \frac{TN}{TN + FP}\right) \qquad (4.23)$$

   where TP is true positive, FP is false positive, TN is true negative, and FN is false negative.

3. Balanced Error Rate[166]

   It is the average of errors on each class. It is also called as *half total error rate*. It is stated as given in Eq. (4.24).

   $$BER = \frac{1}{2}\left(\frac{FP}{P} + \frac{FN}{N}\right) = 1 - BCR \qquad (4.24)$$

   where FP is false positive, FN is false negative, P is total positive, and N is total negative.

JUSTIFICATION OF THIS CHAPTER

In Chapter 3, the method proposed had a basic prerequisite that needed to be fulfilled to make the algorithm work. Support vector machine-Bayesian t-test-recursive feature elimination that was proposed in the previous chapter required a basic calculation of weight vectors that have to be generated from the support vector machine classifiers. This weight vector was one of the foremost criteria for generating the ranking score. Hence, the technique was said to be quite dependable. TR algorithm used in this chapter is an independent technique that does not have any basic prerequisite to be fulfilled.

PROPOSED METHODOLOGIES OF TRACE RATIO ALGORITHM FOR GENE SELECTION AND RANKING

In this chapter, we have proposed two methodologies of using the TR algorithm effectively. In our first method, *IG-TR gene ranking* is proposed that uses IG as the base medium for evaluation along with the original existing TR algorithm, whereas in our second method, *CCA-TR gene ranking* aims at modifying the existing TR algorithm on the basis of scoring criteria.

Method I: Information Gain-Trace Ratio Gene Ranking

In this process, we have kept the original TR algorithm intact. Rather, as a substitute of modifying the algorithm, we distorted the base dataset. This distortion is not abstract but is based on some criteria acknowledgment. That is, after the preprocessing step, the dataset was again rearranged and restructured using the IG value extracted. We calculated IG for the dataset. The higher the IG, the better the information content of the attribute; so we reset the entire dataset based on this attribute content value. It is sorted according to the descending order. Once the dataset is redefined, the TR algorithm is applied over it to rank the genes or attributes. Now, these ranked genes are sorted and passed to the classifier for the purpose of accuracy measurement.

Method II: Canonical Correlation Analysis-Trace Ratio Gene Ranking

There is another method that has been projected for the better performance of the TR algorithm. The TR algorithm usually uses the standard Fisher score or Laplacian score to find the weight matrices. As a replacement for using this form of scoring criterion, we preferred to choose another scoring criterion to replace one of them. For our work, we use the usual Fisher score with CCA factor. That is, our new evaluation criterion for generating the TR or rank of genes is changed from the Fisher score to canonical correlation score. Using this score, we generated the TR score for the genes, which was then passed to a different classifier for the purpose of accuracy estimation. The evaluation criterion for finding the weight matrices or adjacency matrices of the new TR algorithm is stated using Eqs. (4.25) and (4.26).

$$(M_w)_{ij} = \begin{cases} \dfrac{cov(U_i, V_j)}{\sqrt{var(U_i)var(V_i)}}, & \text{if } l_i = l_j \\ 0, & \text{if } l_i \neq l_j \end{cases} \quad (4.25)$$

$$(M_b)_{ij} = \begin{cases} \dfrac{cov(U_i, V_j)}{\sqrt{var(U_i)var(V_i)}}, & \text{if } l_i = l_j \\ 1, & \text{Otherwise} \end{cases} \quad (4.26)$$

where $\dfrac{cov(U_i, V_j)}{\sqrt{var(U_i)var(V_i)}}$ is the canonical correlation score and l_i and l_j denote the class labels.

The detailed restructured algorithm in stated in Algorithm 4.3.

EXPERIMENTATION

In this section, we would embark on the types of datasets used and the basic preprocessing step that is a requisite for the five types of datasets taken for the purpose of normalizing it. This step would be followed by the parametric discussion and measures that have been taken into consideration. We have also presented a schematic view of the proposed model. For the evaluation and analysis, MATLAB® version R2014a was used with the system requirement of 8 GB RAM.

Datasets Used

Expression profiling of colon cancer or colorectal adenomas and normal mucosas from 32 patients was downloaded from the study of Alon et al.[154] This set consists of 31 adenomas and 31 normal mucosas (62 samples) having 2000 genes. To illustrate the molecular developments underlying the alteration of normal colonic epithelium, the transcriptomes of 32 prospectively collected adenomas were equated with those of normal mucosas from the same entities. Similarly, the leukemia dataset was collected from Ref. 147 where the dataset is said to consist of 10,056 genes with 48 samples of both acute lymphocytic leukemia (ALL) and acute myeloid leukemia (AML) (24 ALL and 24 AML each). Apart from these two, few more datasets were taken into consideration such as the medulloblastoma dataset[148] having 5893 genes with 34 samples of 25 C and 9 D, lymphoma dataset[149] having 7070 genes with 77 samples of 58 DLBCL and 19 follicular lymphoma (FL) (Affymetrix HuGeneFL array), and the prostate cancer dataset[150] having 12,533 genes with 102 samples of 50 normal and 52 tumor (Affymetrix Human Genome U95Av2 Array platform). These large-scale gene expression datasets were first measured to be statistically examined and then were used for the assessment of the existing TR algorithm and modified TR algorithm.

Preprocessing

A primary and essential stage of preprocessing is normalization. Normalization process transforms the data into a layout that will be more simply and effectively processed for the purpose of the user. Here, the datasets were normalized using *min-max* normalization.[151] Min-max normalization is an effortless technique where this can particularly fit the data within a predefined boundary. In other words, it is a way that one linearly transforms the real data values such that the minimum and the maximum of the transformed data take certain values. The technique can be represented as shown in Eq. (4.27):

$$x' = \frac{(x - x_{\min})}{(x_{\max} - x_{\min})} \quad (4.27)$$

where $x_{\min}$ is the minimal data value appearing and $x_{\max}$ is the maximal data value appearing.

Parameter Discussion

In earlier section, we have anticipated two different types of methods (*IG-TR gene ranking* and *CCA-TR gene ranking*) for finding the TR of a data matrix. This produces a new rank list of genes, which are then conceded to the variants of artificial neural network (NN) algorithm for the purpose of classification and accuracy measurement. The factors of covariance and variance have been used to provide better results. The top 50, 100, 150, and 200 genes were selected to be used in the TR algorithm for generation of TR and rank list. This assortment was a crucial criterion based on which the entire rank list was generated. IG is also a significant

Step 1:   Calculate adjacency matrices for with-in the class (M_w) and between the classes

(M_b) using Canonical Correlation score as follows (eq (4.25) and eq (4.26)):

$$(M_w)_{ij} = \frac{cov(U_i, V_i)}{\sqrt{var(U_i)*var(V_i)}}, if\, l_i = l_j \wedge 0, Otherwise$$

$$(M_b)_{ij} = \frac{cov(U_i, V_j)}{\sqrt{var(U_i)*var(V_j)}}, if\, l_i = l_j \wedge 1, Otherwise$$

Step 2:   Calculate the diagonal matrices (DM_w and DM_b) for the above adjacency

matricesas given below (as in eq (4.14) and eq (4.15)):

$$(DM_w)_{ii} = \sum_{ij} (M_w)_{ij}$$

$$(DM_b)_{ii} = \sum_{ij}(M_b)_{ij}$$

Step 3:   Calculate Laplacian matrices (LM_w and LM_b) using the eq (4.12) and eq (4.13).

$$LM_w = DM_w - M_w$$

$$LM_b = DM_b - M_b$$

Step 4:   Construct a matrix of k features by initially selecting randomly k features from original dataset (say R_k).

Step 5:   Declare an empty matrix (say N_k) to store top k features after finding scores of each feature .

Step 6:   Repeat steps 6 to 10 until $R_k! = N_k$

Step 7:   Calculate $Y = XLM_b X^T$ and $Z = XLM_w X^T$

Step 8:   Calculate Trace Ratios as $TR_y = TR(R_k^T Y R_k)$ and $TR_z = TR(R_k^T Z R_k)$

Step 9:   Calculate $\beta = \frac{TR_y}{TR_z}$

Step 10: Calculate Score of each feature as $F(f_i) = m_i^T (Y - \beta Z)m_i$

Step 11: Select new top k features based on the score and store in N_k

Step 12: Store final k features R_k for further processing

Step 13: Stop

ALGORITHM 4.3 Modified trace ratio algorithm.

factor that has been considered here for generating high set of genes having huge information content. Hence, the selection of such genes played a major role in finding the TR and rank list. These two processes enhance the chance of finding the appropriate list (rank list) for the purpose of classification to get a better performance value. TR itself is a well-defined algorithm, and merging these extra parameters improves its performance to a higher range. This merger also takes less number of iterations for generating the rank list as compared with the original and unmodified TR algorithm.

Significance and Statistical Analysis of Information Gain

Five datasets are considered that contain a moderately good number of samples and genes. The datasets at the preliminary stage were preprocessed using *min-max normalization*. In the normalized dataset, we used the IG procedure to get hold of the IG vectors for each attribute (i.e., genes), which is then used to sort and reorder the dataset in a descending order. Statistically,

the considerable genes (genes having high amount of IG content) are selected and kept first, and then other genes subsequently follow.

Implementation and Performance Analysis
The proposed schematic model is described in Fig. 4.1.

In the earlier section, we have proposed two methodologies for generating TR. It involved a series of steps where we started with the normalization step that is common to both the methods (*IG-TR gene ranking* and *CCA-TR gene ranking*). Min-max normalization was used with the five datasets for linearly transforming the new information into some specified boundaries. Now, we begin analyzing the first and second methods individually (as shown in method I and method II of Fig. 4.1).

In the first method or *IG-TR gene ranking* of Fig. 4.1, we need to compute the IG of the datasets to find how important a gene vector is. This would additionally be used to select the gene vector or attribute that contains the highest information content. Based on this information content, the data matrix or dataset is sorted and reordered. The reordering is based on the descending

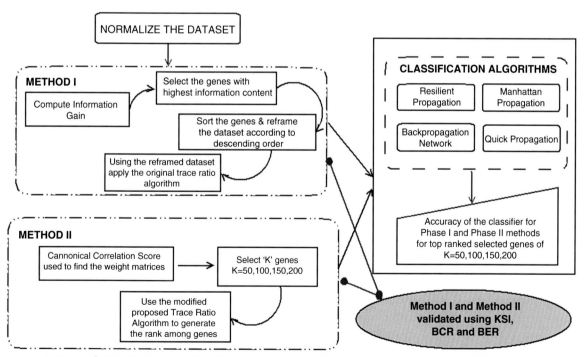

FIG. 4.1 Schematic representation of the proposed model. *BCR*, balanced classification rate; *BER*, balanced error rate; *CCA*, canonical correlation analysis; *IG*, information gain; *KSI*, Kuncheva stability index; *TR*, trace ratio.

order criteria where the gene of high information content is kept first and the genes with least information substance is kept at last. Generally, studies say that attributes of less information substance are not considered important, but in our study every gene vector is given an equal importance and hence kept in the data matrix. Now, this reformed dataset is used as the base input to the TR algorithm, where it was eventually found that with just a modest change in the base data the entire algorithm behaves differently. This difference was mainly on the basis of less number of iterations that was obtained for the convergence of the algorithm (where based on the k number of genes selected, the rank set of all the genes would be same). This difference between the original TR algorithm and modified TR algorithm was also realized through the classification algorithm. The gene rank list was generated and this was given as the input to the classifiers (resilient propagation, backpropagation, quick propagation, and Manhattan propagation), which offers an exceptional accuracy as compared with the original and unmodified TR algorithm.

In the second method or *CCA-TR gene ranking* of Fig. 4.1, we eventually changed the scoring criterion of the TR algorithm instead of changing the base input. As a substitute of Fisher score, we preferred to choose canonical correlation score to determine the weight matrices. CCA is a statistical method that is used to confine the correlation between two variables. Thus, by using the new scoring method, we uncover the TR by selecting the k value (number of genes) as 50, 100, 150, and 200. Now, this modified TR algorithm also generates the rank list with less number of iterations by suitably converging. Now, this new rank list was passed to the variants of the classification algorithm where the classifier's accuracy improved a lot as compared with the original TR algorithm.

Hence, we state that by this modification we established a huge difference in the performance of the TR

algorithm. The difference found at the end provided us with better accuracy value as compared with the unperturbed TR algorithm. Aside from the accuracy measures, the number of iterations required for convergence of the algorithm was eventually less than the original TR algorithm. For most of the datasets, both the methods gave us 100% accuracy or zero error factor.

RESULT ANALYSIS
As stated earlier, five datasets have been considered for the purpose of assessment. Each method's result has been significantly discussed with proper tabular and graphical representations. It was found that both the methods were responding positively and were depicted remarkable results as compared with the original algorithm. The anticipated methods were passed to different classification algorithms, and it was experimental that they varied either in a huge margin or in a smaller margin. Table 4.1 shows the characteristics and features of the dataset considered for the experimental analysis.

Information Gain-Trace Ratio Gene Ranking Algorithm
One of the major issues in gene selection progression is that with a huge ordered gene subset, we need to find such a set (reordered set) where the classification accuracy would be higher. Based on this concept of gene selection, we tried to reform the dataset or data matrix into a proper ranked set and then this set was compared with the original dataset. In other words, a clear comparison was drawn by considering the original preprocessed dataset and reframed dataset that was passed as the base input to the existing TR algorithm. This was further materialized and validated by different classification algorithms. The results are shown in Tables 4.2—4.6 where Table 4.6 shows the average

TABLE 4.1
Description of the Datasets Used in Experimental Analysis

Data	Amount of Genes	Number of Samples		Training Data	Testing Data	References
		Class 1	Class 2			
Colon cancer	2000	31 (N)	31 (T)	44	18	154
Leukemia	10,056	24 (ALL)	24 (AML)	34	14	147
Medulloblastoma	5893	25(C)	9 (D)	24	10	148
Lymphoma	7070	58 (D)	19 (FL)	54	23	149
Prostate cancer	12,533	50 (N)	52 (T)	44	18	150

ALL, acute lymphocytic leukemia; *AML*, acute myeloid leukemia; *FL*, follicular lymphoma.

TABLE 4.2
Performance Assessment of Reframed Input With Original Preprocessed Base Input Where k = 50 for Trace Ratio Algorithm

Dataset	Resilient Propagation				Quick Propagation				Backpropagation				Manhattan Propagation			
	Original		Infogain		Original		Infogain		Original		Infogain		Original		Infogain	
	Acc	Itr	Acc	Itr	Acc	Itr	Acc	Itr	Acc	Itr	Acc	Itr	Acc	Itr	Acc	Itr
Colon cancer	98.36	122	100	111	98.22	952	100	840	79.03	2024	100	628	80.78	2145	87.45	2043
Leukemia	91.6	12	100	12	90.24	417	100	390	91.66	14	100	13	89.47	321	92.5	224
Medulloblastoma	99.10	44	100	35	98.72	655	100	564	76.47	725	100	602	79.65	856	85	601
Lymphoma	98.70	2335	98.70	1347	97	4356	99.01	4210	97.40	4901	100	3568	95.23	3568	97.23	2864
Prostate cancer	99.01	8197	100	983	98.23	4055	100	3910	50.98	4987	71.28	4029	65.25	3687	76	3402

Acc, accuracy; Itr, iteration.

TABLE 4.3
Performance Assessment of Reframed Input With Original Preprocessed Base Input Where k = 100 for Trace Ratio Algorithm

Dataset	Resilient Propagation				Quick Propagation				Backpropagation				Manhattan Propagation			
	Original		Infogain		Original		Infogain		Original		Infogain		Original		Infogain	
	Acc	Itr	Acc	Itr	Acc	Itr	Acc	Itr	Acc	Itr	Acc	Itr	Acc	Itr	Acc	Itr
Colon cancer	98.3	108	100	99	87.23	874	90.45	754	97.24	1042	98.3	967	82	2256	88	1568
Leukemia	100	16	100	12	90	547	95	489	98.45	25	100	9	92.01	426	95.24	368
Medulloblastoma	99.01	41	100	38	92.86	702	94	547	98.24	3988	99.01	4011	86.35	965	88.25	867
Lymphoma	98.70	1452	98.70	1321	97.58	4221	98	3587	99	4254	100	3987	92.48	4781	94.23	3892
Prostate cancer	99.01	4496	100	1337	89.47	4068	95.68	3985	95.25	2471	99.01	1761	72	3874	82.89	3471

Acc, accuracy; Itr, iteration.

TABLE 4.4
Performance Assessment of Reframed Input With Original Preprocessed Base Input Where $k = 150$ for Trace Ratio Algorithm

Dataset	Resilient Propagation Original		Infogain		Quick Propagation Original		Infogain		Backpropagation Original		Infogain		Manhattan Propagation Original		Infogain	
	Acc	Itr	Acc	Itr	Acc	Itr	Acc	Itr	Acc	Itr	Acc	Itr	Acc	Itr	Acc	Itr
Colon cancer	99.01	123	100	89	88.56	954	93.41	842	97.58	1234	99.25	1047	86.47	2458	95.25	2110
Leukemia	91.6	12	100	12	91	658	93.42	524	93.57	38	98.20	23	93.78	528	100	403
Medulloblastoma	99.01	56	100	30	92.58	834	95.41	712	98.65	4078	100	3854	89.41	913	93.58	804
Lymphoma	98.70	2190	98.70	1576	93.58	4025	94.14	3651	93.47	4378	96.58	3821	91.47	4582	93.58	4036
Prostate cancer	99.01	4108	100	1128	86.78	3958	89.45	3822	88.25	2854	92.42	2103	78.58	3451	85.34	3241

Acc, accuracy; Itr, iteration.

TABLE 4.5
Performance Assessment of Reframed Input With Original Preprocessed Base Input Where $k = 200$ for Trace Ratio Algorithm

Dataset	Resilient Propagation Original		Infogain		Quick Propagation Original		Infogain		Backpropagation Original		Infogain		Manhattan Propagation Original		Infogain	
	Acc	Itr	Acc	Itr	Acc	Itr	Acc	Itr	Acc	Itr	Acc	Itr	Acc	Itr	Acc	Itr
Colon cancer	98.15	145	100	105	91.69	1025	96.24	852	95.85	1354	97.25	1204	88	2147	93.45	1543
Leukemia	100	16	100	14	90.47	705	92.35	642	92.01	42	95.21	33	91.47	682	95.47	541
Medulloblastoma	98.36	51	100	32	93.45	920	96.24	824	97.36	3954	100	3241	92.54	1054	97.48	932
Lymphoma	98.70	1304	100	1730	94.25	3982	96.24	3475	92.58	3256	96.24	2543	90.47	4421	95.3	4102
Prostate cancer	99.01	4482	100	1575	88	3841	90.24	3654	86.21	2745	90.01	2235	80.25	3647	85.78	3278

Acc, accuracy; Itr, iteration.

TABLE 4.6
Average Performance Assessment of Reframed Input With Original Preprocessed Base Input Where $k = 500$ for Trace Ratio Algorithm for 10 Runs

	Resilient Propagation				Quick Propagation				Backpropagation				Manhattan Propagation			
	Original		Infogain		Original		Infogain		Original		Infogain		Original		Infogain	
Dataset	Acc	Itr	Acc	Itr	Acc	Itr	Acc	Itr	Acc	Itr	Acc	Itr	Acc	Itr	Acc	Itr
Colon cancer	98.21	132	100	124	92.35	854	94.78	742	94.25	1249	96.87	1145	89	1785	95	1547
Leukemia	97.9	13	100	13	90.42	650	94.57	521	93.78	65	96.12	51	92.56	742	94.25	547
Medulloblastoma	99.02	45	100	28	90.01	1234	96.85	1074	96.98	3675	98.36	3085	91.54	985	94.56	784
Lymphoma	98.7	1110	99.48	1537	93.01	3724	95.21	3612	90.14	3412	93.85	3200	88.45	4085	89	3547
Prostate cancer	99.11	3713	99.60	1038	86.45	2564	88.47	2410	89.56	2450	93.85	2286	78.62	3742	82.54	3410

Acc, accuracy; Itr, iteration.

accuracy obtained with 10 runs of iterations. It was observed that from Tables 4.2−4.6 the proposed first method provided a good result in terms of accuracy with different classifiers.

Canonical Correlation Analysis-Trace Ratio Gene Ranking Algorithm

In this method we have evaluated the results of the existing TR algorithm and proposed TR algorithm on the five datasets selected. The results were again validated on different classifiers, and they are shown in Tables 4.7−4.11 where Table 4.11 depicts the average accuracy obtained with 10 runs. It was observed that from Tables 4.7−4.11 the accuracy obtained for the proposed second method quite appreciates for the different NN classifier variants.

Performance Assessment of Method I and Method II (Information Gain-Trace Ratio and Canonical Correlation Analysis-Trace Ratio Gene Ranking Algorithm)

To assess the proposed methods, three different types of metrics for evaluation were considered for the five datasets taken. Although there exist several performance indexes and metrics, KSI, BCR, and BER are the three metrics that are chosen for the assessment of the methods. It was observed that the proposed methods provided suitable results as compared with the original and unmodified algorithm. Tables 4.12−4.16 depict the results obtained by the metric evaluation for five different types of datasets: colon cancer, leukemia, medulloblastoma, lymphoma, and prostate cancer.

It is evident that from Tables 4.12−4.16 the proposed two methods are providing quite satisfying outcomes as compared with the original algorithm. For $k = 50$, 100, 150, and 200, results provided for all the datasets are satisfying and quite encouraging. In certain cases, it has been observed that the results in both the proposed methods are not much different, whereas there are cases in which one of the two methods is providing a better result. For KSI, it was found that the result approached to 1 (where intersection between two subsets is more, i.e., good gene subset selection) than 0 (where intersection between two subsets is 0, i.e., bad gene subset selection). In our algorithm, the rank of the gene changes at each iteration, and finally in the last iteration the rank is same as the k randomly selected data. Hence, an average of all iterations is considered except the last one. In other words, if there are n number of iterations, taking place in the TR algorithm (for both the existing and the proposed algorithms), we will consider an average of $(n - 1)$

iteration's KSI. We are intentionally leaving the last iteration, as the intersection of the subsets will lead us to a KSI value of 1. As a result of this, the KSI value would be more biased toward 1. The second metric BCR is considered as another parameter for evaluating the methods proposed. Here, the more the result approaches to 1, the better the classification rate and the better the selection of genes. The third metric is BER that can be obtained by performing the evaluation of 1-BCR, and the more the result approaches to 0, the better the evaluation of the gene subset selection. In other words, the error rate in any method proposed should be less, i.e., approaching to 0. Moreover, for varying value of k, there is quite a few fluctuation in the result, and it can be stated that with less amount of k chosen, we are getting better results as compared with higher k value.

DISCUSSION

This work can be finally summarized as follows:

- To start with, the dataset has been normalized using min-max normalization process to which the original and existing TR algorithm can be used.
- From the existing TR algorithm, the rank of the genes was extracted and the dataset was reformed according to the new rank generated. Now, this dataset was conceded to a classifier and the accuracy of the dataset was assessed.
- As TR algorithm is a powerful ranking technique, we thought of slightly restructuring it. Two different approaches were considered, and hence two new methods, namely, *IG-TR gene ranking* and *CCA-TR gene ranking*, were proposed.
- In the first method, from the datasets IG was computed and the genes containing the highest information content were selected. Based on this selection, the datasets were redesigned in the descending order. The existing TR algorithm was now used on these redesigned datasets, and ranks were generated (by considering random amount of k value). The newly generated ranked set was passed to the classifier (variants of NN), and its accuracy was tested. It was found that by slightly changing the input pattern of the TR algorithm, the accuracy obtained was quite high.
- To add a new instance to this finding, another method was proposed where the dataset was intact; rather a whole new TR algorithm's scoring or ranking method was proposed.
- Instead of considering Fisher score as the criterion for scoring or ranking in the traditional TR algorithm, we selected another statistical technique

TABLE 4.7
Performance Assessment of the Existing TR Algorithm and Proposed TR Algorithm Where k = 50 for the TR Algorithm

	Resilient Propagation				Quick Propagation				Backpropagation				Manhattan Propagation			
	Original		CCA-TR		Original		CCA-TR		Original		CCA-TR		Original		CCA-TR	
Dataset	Acc	Itr	Acc	Itr	Acc	Itr	Acc	Itr	Acc	Itr	Acc	Itr	Acc	Itr	Acc	Itr
Colon cancer	98.36	122	**100**	**133**	87.23	874	**89.85**	**648**	79.03	2024	**83.58**	**1524**	80.78	2145	**89.55**	**2048**
Leukemia	91.6	12	**100**	**11**	90	547	**98.14**	**347**	91.66	14	**94.88**	**14**	89.47	321	**92.66**	**256**
Medulloblastoma	99.10	44	**100**	**29**	92.86	702	**96.31**	**587**	76.47	725	**85.86**	**540**	79.65	856	**86.34**	**785**
Lymphoma	98.70	2335	**100**	**604**	97.58	4221	**98.25**	**3511**	97.40	4901	**98.11**	**3124**	95.23	3568	**98.35**	**2475**
Prostate cancer	99.01	8197	**99.01**	**1777**	89.47	4068	**90**	**3314**	50.98	4987	**86.47**	**2589**	65.25	3687	**79.36**	**3549**

Acc, accuracy; CCA, canonical correlation analysis; Itr, iteration; TR, trace ratio.

TABLE 4.8
Performance Assessment of the Existing TR Algorithm and Proposed TR Algorithm Where k = 100 for the TR Algorithm

	Resilient Propagation				Quick Propagation				Backpropagation				Manhattan Propagation			
	Original		CCA-TR		Original		CCA-TR		Original		CCA-TR		Original		CCA-TR	
Dataset	Acc	Itr	Acc	Itr	Acc	Itr	Acc	Itr	Acc	Itr	Acc	Itr	Acc	Itr	Acc	Itr
Colon cancer	98.3	108	**100**	**93**	87.23	874	**90.24**	**810**	97.24	1042	**99.04**	**856**	82	2256	**88.33**	**1689**
Leukemia	100	16	**100**	**26**	90	547	**96.33**	**321**	98.45	25	**100**	**20**	92.01	426	**94.56**	**354**
Medulloblastoma	99.01	41	**100**	**29**	92.86	702	**98.54**	**652**	98.24	3988	**100**	**3865**	86.35	965	**90.01**	**892**
Lymphoma	98.70	1452	**98.7**	**764**	97.58	4221	**99.33**	**3214**	99	4254	**99.58**	**3358**	92.48	4781	**97.68**	**3658**
Prostate cancer	99.01	4496	**99.01**	**921**	89.47	4068	**96.8**	**2549**	95.25	2471	**99.41**	**2045**	72	3874	**80.56**	**2546**

Acc, accuracy; CCA, canonical correlation analysis; Itr, iteration; TR, trace ratio.

TABLE 4.9
Performance Assessment of the Existing TR Algorithm and Proposed TR Algorithm Where $k = 150$ for the TR Algorithm

| Dataset | Resilient Propagation | | | | Quick Propagation | | | | Backpropagation | | | | Manhattan Propagation | | | |
| | Original | | CCA-TR | | Original | | CCA-TR | | Original | | CCA-TR | | Original | | CCA-TR | |
	Acc	Itr	Acc	Itr	Acc	Itr	Acc	Itr	Acc	Itr	Acc	Itr	Acc	Itr	Acc	Itr
Colon cancer	99.01	123	100	79	88.56	954	97	852	97.58	1234	100	1025	86.47	2458	96.87	2053
Leukemia	91.6	12	100	21	91	658	96.78	542	93.57	38	98.69	29	93.78	528	97.25	423
Medulloblastoma	99.01	56	100	24	92.58	834	98.32	724	98.65	4078	100	3569	89.41	913	95.21	821
Lymphoma	98.70	2190	98.7	828	93.58	4025	99	4015	93.47	4378	98.77	3698	91.47	4582	96.22	3548
Prostate cancer	99.01	4108	99.01	608	86.78	3958	90.25	3822	88.25	2854	93.20	2264	78.58	3451	83.27	3025

Acc, accuracy; CCA, canonical correlation analysis; Itr, iteration; TR, trace ratio.

TABLE 4.10
Performance Assessment of the Existing TR Algorithm and Proposed TR Algorithm Where $k = 200$ for the TR Algorithm

| Dataset | Resilient Propagation | | | | Quick Propagation | | | | Backpropagation | | | | Manhattan Propagation | | | |
| | Original | | CCA-TR | | Original | | CCA-TR | | Original | | CCA-TR | | Original | | CCA-TR | |
	Acc	Itr	Acc	Itr	Acc	Itr	Acc	Itr	Acc	Itr	Acc	Itr	Acc	Itr	Acc	Itr
Colon cancer	98.15	145	100	88	91.69	1025	99.05	1015	95.85	1354	98.65	1023	88	2147	92.35	2014
Leukemia	100	16	100	23	90.47	705	95.87	653	92.01	42	96.35	32	91.47	682	96.85	586
Medulloblastoma	98.36	51	100	38	93.45	920	100	854	97.36	3954	100	2542	92.54	1054	98.31	993
Lymphoma	98.70	1304	98.7	669	94.25	3982	98.35	3821	92.58	3256	96.87	3105	90.47	4421	93.54	3214
Prostate cancer	99.01	4482	99.01	515	88	3841	92.34	3542	86.21	2745	92.85	2105	80.25	3647	89.32	2598

Acc, accuracy; CCA, canonical correlation analysis; Itr, iteration; TR, trace ratio.

TABLE 4.11
Average Performance Assessment of the Existing TR Algorithm and Proposed TR Algorithm Where $k = 500$ for the TR Algorithm With 10 Runs

Dataset	Resilient Propagation				Quick Propagation				Backpropagation				Manhattan Propagation			
	Original		CCA-TR		Original		CCA-TR		Original		CCA-TR		Original		CCA-TR	
	Acc	Itr	Acc	Itr	Acc	Itr	Acc	Itr	Acc	Itr	Acc	Itr	Acc	Itr	Acc	Itr
Colon cancer	98.21	132	100	70	92.35	854	99.05	781	94.25	1249	99.36	1156	89	1785	93.68	1622
Leukemia	97.9	13	100	13	90.42	650	96.87	520	93.78	65	98.32	52	92.56	742	97.85	689
Medulloblastoma	99.02	45	99.70	28	90.01	1234	96.88	1203	96.98	3675	99	3458	91.54	985	95.84	862
Lymphoma	98.7	1110	99.61	1056	93.01	3724	97.41	3654	90.14	3412	95.32	2596	88.45	4085	92.54	3845
Prostate cancer	99.11	3713	99.31	638	86.45	2564	91.24	2241	89.56	2450	94.58	2150	78.62	3742	83.57	3548

Acc, accuracy; CCA, canonical correlation analysis; Itr, iteration; TR, trace ratio.

TABLE 4.12
Performance Assessment for Colon Cancer Dataset

k Value	Index Metrics	Resilient Propagation			Quick Propagation			Backpropagation			Manhattan Propagation		
		Original	IG-TR	CCA-TR	Original	IG-TR	CCA-TR	Original	IG-TR	CCA-TR	Original	IG-TR	CCA-TR
50	KSI	0.26	0.62	0.72	0.23	0.52	0.75	0.19	0.55	0.69	0.25	0.46	0.55
	BCR	0.96	0.99	0.99	0.97	0.99	0.83	0.76	0.99	0.81	0.75	0.83	0.86
	BER	0.04	0.01	0.01	0.03	0.01	0.17	0.24	0.01	0.19	0.25	0.17	0.14
100	KSI	0.28	0.57	0.69	0.21	0.55	0.62	0.24	0.52	0.70	0.28	0.44	0.53
	BCR	0.93	0.99	0.99	0.85	0.86	0.88	0.93	0.96	0.97	0.77	0.84	0.81
	BER	0.07	0.01	0.01	0.15	0.14	0.12	0.07	0.04	0.03	0.23	0.16	0.19
150	KSI	0.25	0.49	0.62	0.25	0.51	0.63	0.22	0.50	0.65	0.23	0.39	0.54
	BCR	0.94	0.99	0.99	0.83	0.91	0.94	0.92	0.98	0.99	0.82	0.93	0.94
	BER	0.06	0.01	0.01	0.17	0.09	0.06	0.08	0.02	0.01	0.18	0.07	0.06
200	KSI	0.25	0.51	0.63	0.29	0.49	0.60	0.24	0.45	0.66	0.25	0.47	0.51
	BCR	0.94	0.99	0.99	0.88	0.93	0.97	0.92	0.94	0.96	0.86	0.88	0.87
	BER	0.06	0.01	0.01	0.12	0.07	0.03	0.08	0.06	0.04	0.14	0.12	0.13

BCR, balanced classification rate; BER, balanced error rate; CCA, canonical correlation analysis; IG, information gain; KSI, Kuncheva stability index; TR, trace ratio.

TABLE 4.13
Performance Assessment for Leukemia Dataset

k Value	Index Metrics	Resilient Propagation			Quick Propagation			Backpropagation			Manhattan Propagation		
		Original	IG-TR	CCA-TR	Original	IG-TR	CCA-TR	Original	IG-TR	CCA-TR	Original	IG-TR	CCA-TR
50	KSI	0.30	0.57	0.63	0.31	0.49	0.61	0.25	0.42	0.60	0.30	0.44	0.51
	BCR	0.93	0.99	0.99	0.89	0.99	0.95	0.87	0.99	0.91	0.76	0.89	0.88
	BER	0.07	0.01	0.01	0.11	0.01	0.05	0.13	0.01	0.09	0.24	0.11	0.12
100	KSI	0.26	0.46	0.62	0.31	0.49	0.60	0.21	0.39	0.52	0.32	0.45	0.49
	BCR	0.97	0.99	0.99	0.83	0.93	0.92	0.93	0.99	0.99	0.89	0.93	0.90
	BER	0.03	0.01	0.01	0.17	0.07	0.08	0.07	0.01	0.01	0.11	0.07	0.10
150	KSI	0.22	0.47	0.65	0.30	0.42	0.55	0.15	0.36	0.53	0.27	0.44	0.45
	BCR	0.89	0.99	0.99	0.87	0.91	0.92	0.91	0.97	0.96	0.90	0.99	0.91
	BER	0.11	0.01	0.01	0.13	0.09	0.08	0.09	0.03	0.04	0.10	0.01	0.09
200	KSI	0.19	0.48	0.62	0.33	0.41	0.52	0.18	0.33	0.49	0.29	0.48	0.50
	BCR	0.99	0.99	0.99	0.88	0.88	0.85	0.89	0.90	0.92	0.83	0.90	0.90
	BER	0.01	0.01	0.01	0.12	0.12	0.15	0.11	0.10	0.08	0.17	0.10	0.10

BCR, balanced classification rate; BER, balanced error rate; CCA, canonical correlation analysis; IG, information gain; KSI, Kuncheva stability index; TR, trace ratio.

TABLE 4.14
Performance Assessment for Medulloblastoma Dataset

k Value	Index Metrics	Resilient Propagation			Quick Propagation			Backpropagation			Manhattan Propagation		
		Original	IG-TR	CCA-TR	Original	IG-TR	CCA-TR	Original	IG-TR	CCA-TR	Original	IG-TR	CCA-TR
50	KSI	0.22	0.61	0.69	0.23	0.54	0.70	0.22	0.57	0.72	0.23	0.42	0.68
	BCR	0.98	0.99	0.99	0.98	0.99	0.91	0.73	0.99	0.82	0.62	0.81	0.82
	BER	0.02	0.01	0.01	0.02	0.01	0.09	0.27	0.01	0.18	0.38	0.19	0.18
100	KSI	0.23	0.53	0.70	0.25	0.52	0.68	0.25	0.55	0.55	0.29	0.36	0.66
	BCR	0.94	0.99	0.99	0.88	0.89	0.96	0.95	0.98	0.99	0.83	0.86	0.86
	BER	0.06	0.01	0.01	0.12	0.11	0.04	0.05	0.02	0.01	0.17	0.14	0.14
150	KSI	0.27	0.55	0.72	0.22	0.56	0.69	0.21	0.49	0.52	0.28	0.31	0.62
	BCR	0.95	0.99	0.99	0.90	0.90	0.96	0.94	0.99	0.99	0.86	0.90	0.92
	BER	0.05	0.01	0.01	0.10	0.10	0.04	0.06	0.01	0.01	0.14	0.10	0.08
200	KSI	0.24	0.53	0.70	0.27	0.48	0.70	0.22	0.42	0.49	0.26	0.32	0.56
	BCR	0.96	0.99	0.99	0.92	0.91	0.99	0.96	0.99	0.99	0.88	0.93	0.99
	BER	0.04	0.01	0.01	0.08	0.09	0.01	0.04	0.01	0.01	0.12	0.07	0.01

BCR, balanced classification rate; BER, balanced error rate; CCA, canonical correlation analysis; IG, information gain; KSI, Kuncheva stability index; TR, trace ratio.

TABLE 4.15
Performance Assessment for Lymphoma Dataset

k Value	Index Metrics	Resilient Propagation			Quick Propagation			Backpropagation			Manhattan Propagation		
		Original	IG-TR	CCA-TR	Original	IG-TR	CCA-TR	Original	IG-TR	CCA-TR	Original	IG-TR	CCA-TR
50	KSI	0.29	0.55	0.68	0.22	0.48	0.65	0.18	0.47	0.71	0.22	0.42	0.55
	BCR	0.99	0.97	0.99	0.95	0.98	0.94	0.95	0.99	0.96	0.92	0.95	0.96
	BER	0.01	0.03	0.01	0.05	0.02	0.06	0.05	0.01	0.04	0.08	0.05	0.04
100	KSI	0.31	0.52	0.72	0.27	0.48	0.70	0.25	0.51	0.70	0.21	0.45	0.54
	BCR	0.94	0.97	0.96	0.93	0.96	0.97	0.97	0.99	0.97	0.90	0.91	0.93
	BER	0.06	0.03	0.04	0.04	0.04	0.03	0.03	0.01	0.03	0.10	0.09	0.07
150	KSI	0.27	0.55	0.71	0.25	0.52	0.65	0.27	0.55	0.68	0.22	0.49	0.52
	BCR	0.95	0.95	0.94	0.88	0.91	0.97	0.91	0.92	0.97	0.86	0.91	0.94
	BER	0.05	0.05	0.06	0.12	0.09	0.03	0.09	0.08	0.03	0.14	0.09	0.06
200	KSI	0.26	0.60	0.77	0.26	0.56	0.55	0.25	0.50	0.71	0.16	0.47	0.44
	BCR	0.97	0.99	0.94	0.91	0.92	0.90	0.91	0.92	0.92	0.85	0.94	0.89
	BER	0.03	0.01	0.06	0.09	0.08	0.10	0.09	0.08	0.08	0.15	0.06	0.11

BCR, balanced classification rate; BER, balanced error rate; CCA, canonical correlation analysis; IG, information gain; KSI, Kuncheva stability index; TR, trace ratio.

TABLE 4.16
Performance Assessment for Prostate Cancer Dataset

k Value	Index Metrics	Resilient Propagation			Quick Propagation			Backpropagation			Manhattan Propagation		
		Original	IG-TR	CCA-TR	Original	IG-TR	CCA-TR	Original	IG-TR	CCA-TR	Original	IG-TR	CCA-TR
50	KSI	0.13	0.36	0.52	0.15	0.22	0.49	0.20	0.34	0.52	0.24	0.28	0.48
	BCR	0.95	0.99	0.97	0.96	0.98	0.87	0.48	0.69	0.82	0.61	0.74	0.72
	BER	0.05	0.01	0.03	0.04	0.02	0.13	0.52	0.31	0.18	0.39	0.26	0.28
100	KSI	0.20	0.42	0.55	0.18	0.30	0.51	0.20	0.31	0.48	0.21	0.32	0.42
	BCR	0.96	0.99	0.97	0.86	0.92	0.92	0.91	0.97	0.97	0.68	0.81	0.75
	BER	0.04	0.01	0.03	0.14	0.08	0.08	0.09	0.03	0.03	0.32	0.19	0.25
150	KSI	0.21	0.50	0.58	0.20	0.33	0.45	0.19	0.34	0.50	0.19	0.32	0.47
	BCR	0.97	0.99	0.97	0.81	0.85	0.88	0.84	0.88	0.89	0.72	0.82	0.79
	BER	0.03	0.01	0.03	0.11	0.15	0.12	0.16	0.12	0.11	0.28	0.18	0.21
200	KSI	0.22	0.55	0.51	0.22	0.32	0.49	0.16	0.32	0.45	0.10	0.33	0.42
	BCR	0.96	0.99	0.97	0.82	0.86	0.82	0.83	0.85	0.89	0.74	0.82	0.85
	BER	0.04	0.01	0.03	0.18	0.14	0.18	0.17	0.15	0.11	0.26	0.18	0.15

BCR, balanced classification rate; *BER*, balanced error rate; *CCA*, canonical correlation analysis; *IG*, information gain; *KSI*, Kuncheva stability index; *TR*, trace ratio.

called CCA score for the generation of new rank list or set. It was observed that the rank generated out of this technique and the new dataset formed produced a better classification accuracy as compared with the traditional TR algorithm.

- Finally, the two methods, although provided better results in comparison with the traditional TR algorithm, need to be validated and assessed properly. Hence, KSI, BCR, and BER were the three performance metrics chosen for assessing their performance. Lastly, it was proved that the two proposed techniques provided better validation results as compared with the original algorithm.

SUMMARY

In this chapter, two methods *IG-TR gene ranking* and *CCA-TR gene ranking* were proposed. These methods were assessed with five types of datasets. The basic aim of the proposed techniques was to rank the genes with few numbers of randomly selected genes. It was proved that the ranks generated out of these techniques were quite good, and they were further validated by passing them through a classification stage. The accuracy of the classifiers obtained provided a suitable and valid means to assess the two techniques proposed. For our work, four variants of NN classifiers (resilient propagation, quick propagation, backpropagation, and Manhattan propagation) were selected; however, we can also choose any other classification technique for the purpose of measuring classification accuracy. It was observed that the accuracy of all the classifiers is more or less the same, but the proposed method's accuracy was far better as compared with the existing algorithm. For rank generation, different k values are considered and it was concluded that by choosing a small set of k, we are able to acquire a better ranking pattern for the genes. The two methods along with the existing method are further validated using different performance metrics such as KSI, BCR, and BER. Ultimately, the genes obtained from these methods can be used for the purpose of constructing GRNs. In the next chapter, another gene selection technique that improves the hefty calculation involved in this method to a simplified process of calculation has been processed. Signal-to-noise ratio is used along with the TR algorithm for the purpose of widespread gene selection process.

CHAPTER 5

SNR-TR Gene Ranking Method: A Signal-to-Noise Ratio–Based Gene Selection Algorithm Using Trace Ratio for Gene Expression Data

SHRUTI MISHRA, PhD • SANDEEP KUMAR SATAPATHY, PhD • DEBAHUTI MISHRA, PhD

INTRODUCTION

Gene regulatory networks (GRNs) play a major role in calculating the biologic process and molecular organization of living organisms.[128,129] The only major fear that always exists is the modeling of the networks. Several computational approaches have been sprung up for accessing the gene expression and finding the regulator network and components.[131] However, in the current scenario, developing the GRN model is a major challenge in the biologic research.[132] GRN models provide an insight view about the observational facts of the data or the genes, their interaction patterns between the data, and ultimately factors affecting the interactions. In other words, the model allows us to deliver an overall dynamic behavior of the network.

To develop a GRN model, a reverse engineering path can be followed where instead of creating the network from the whole genome sequence or gene expression data, we can establish a small subset of informative genes from which the regulatory interaction pattern can be generated.[133] The formation of these small subsets of genes is known as gene selection.[134] Gene selection from microarray data is a highly statistically significant problem that has persisted for long time. One of the major issues is the number of samples that are usually less as compared with the thousands of genes whose expression levels are measured.[134]

We know that from thousands of genes available in the gene expression data, all genes are not relevant and disease causing. Hence, it is every bit important to extract those relevant and diseases-causing genes from the vast set of irrelevant data available.[132] One of the methods for obtaining the above is through selection or ranking. Gene selection or feature selection can be defined using three fundamental methods or categories: filter method, wrapper method, and embedded method.[135–137] Filter method can be obtained using two operations such as ranking and subset selection based on the ranking, i.e., it assigns scores to each feature.[139] Wrapper methods use the search problem where different combinations of the features are prepared, extracted, and compared.[140] Embedded methods select the feature that contributes more in the accuracy of any model that is created.[141]

In this chapter, the scoring method for the existing trace ratio (TR) algorithm has been redesigned using the signal-to-noise ratio (SNR) scoring technique. The existing TR algorithm is mainly based on the Fisher score or Laplacian score that we replaced with a new scoring scheme for calculation of the weight matrices within the class and between the class. The proposed method was then assessed with five benchmark datasets such as colon cancer, leukemia, medulloblastoma, lymphoma, and prostate cancer. The dataset is quite large enough in terms of number of genes but has minimal number of sample size. When compared with the existing TR algorithm, it was found that the SNR-TR algorithm outperformed in terms of classification accuracy and number of iterations. It was established that the two classifiers (resilient propagation and backpropagation) that were considered provided excellent results in the SNR-TR gene ranking than the TR algorithm, with less number of iterations. The result was compared with different k values such as 50, 100, and 150 where k is the number of genes selected initially and ultimately.

PRELIMINARIES

In this section, some preliminary concepts and facts have been stated for the techniques SNR and TR (discussed in Trace Ratio section). Apart from these said techniques, some of the performance indexes such as Kuncheva stability index (KSI), balanced classification rate (BCR), and balanced error rate (BER) have been mentioned and defined in Performance Metrices section.

Signal-to-Noise Ratio

One of the most widely used techniques is SNR.[122,123] It is quite popular because of its simplicity and ease of using it. The quality of the biologic data can be assessed as the amount of biologic signal to the amount of noise. The best part in the SNR technique is it depends not only on the noise factor but also on the amount of signal. The SNR is also stated as a statistical measure for finding the effectiveness of the feature in identifying a class out of another class. In other words, it identifies the pattern with a minimal difference in mean expression between two groups used and a minimal variation of the expression within each group. The description of the score can be stated as in Eq. (5.1):

$$\text{SNR} = \left| \frac{\mu_1 - \mu_2}{\sigma_1 + \sigma_2} \right| \qquad (5.1)$$

where μ_1 and μ_2 denote the means of the expression level for the samples in class 1 and class 2, respectively. σ_1 and σ_2 denote the standard deviations for the samples in each class. The SNR technique helps to select the features that have high SNR, i.e., maximal SNR (or minimal error).

JUSTIFICATION OF THIS CHAPTER

In Chapter 4, information gain-trace ratio (IG-TR) and canonical correlation analysis-trace ratio (CCA-TR) gene ranking algorithms were used. Both these techniques had their relative limitations such as in IG-TR gene ranking algorithm the dataset had to be reframed and then that reframed dataset was used with the original TR algorithm. This was time-consuming and difficult, as the dataset needed to be particularly sorted. In case of the CCA-TR gene ranking algorithm, the CCA technique needed calculation of covariance and variance for sample-wise data. This is quite hefty and tedious in the aspect of calculation. Hence, to resolve the above said issues, the SNR-TR gene ranking algorithm was proposed, as it involved no reordering of the dataset and no hefty calculation. It only involved simple calculations such as mean and standard deviation.

PROPOSED SIGNAL-TO-NOISE RATIO-TRACE RATIO GENE RANKING ALGORITHM

Here, a thorough discussion regarding the proposed SNR-TR gene ranking algorithm is settled that uses SNR and TR algorithms as the basic parameters. Here, the SNR is used as a new scoring method that replaces the original scoring method of the TR algorithm. The original scoring method that the TR algorithm uses is usually Fisher score or Laplacian score. Here, this generalized scoring method is replaced by the proposed SNR scoring method, which promises to provide a better result as compared with the original TR algorithm. The new developed rank for adjacency matrices within the class (M_w) and between the class (M_b) using the SNR is shown as in Eqs. (5.2) and (5.3).

$$(M_w)_{ij} = \left| \frac{\mu_i - \mu_j}{\sigma_i + \sigma_j} \right|, \text{if } l_i = l_j \wedge 0, \ Otherwise \qquad (5.2)$$

$$(M_b)_{ij} = \left| \frac{\mu_i - \mu_j}{\sigma_i + \sigma_j} \right|, \text{if } l_i \neq l_j \wedge 0, \ Otherwise \qquad (5.3)$$

The detailed SNR-TR gene ranking algorithm is mentioned in Algorithm 5.1.

EXPERIMENTATION

In this section, we present the basic preprocessing step that was used for our five datasets. This would be further followed by the types of datasets used. For the evaluation process, MATLAB® version R2014a was used with the system requirement of 8 GB RAM.

Datasets Used

The datasets used in this section are same as in Experimentation section, Table 4.1.

Preprocessing

The preprocessing technique used in this section is same as in Experimentation section.

Parameter Discussion

In this section, the SNR-TR gene ranking algorithm has been used for finding the TR of a dataset, using which genes can be ranked. In other words, using this proposed gene ranking algorithm, a new rank list is generated that is further passed to the variants of neural network (NN) (resilient propagation and backpropagation) for classification accuracy measurement. The concept has been simplified using the mean and standard deviation for adjacency matrix rank relation. For the generation of TR and TR-based rank list, we selected top 50, 100, and 150 genes. Besides these, top 200 genes were considered and five runs of the algorithm

Step 1:   Calculate adjacency matrices for with in the class (M_w) and between the classes (M_b) using SNR as follows

$$(M_w)_{ij} = \left|\frac{\mu_i - \mu_j}{\sigma_i + \sigma_j}\right|, if\ l_i = l_j \wedge 0, Otherwise$$

$$(M_b)_{ij} = \left|\frac{\mu_i - \mu_j}{\sigma_i + \sigma_j}\right|, if\ l_i \neq l_j \wedge 0, Otherwise$$

where, i and j are sample numbers.

Step 2:   Calculate the diagonal matrices (DM_w and DM_b) for the above adjacency matrices as given below

$$(DM_w)_{ii} = \sum_{ij}(M_w)_{ij}$$
$$(DM_b)_{ii} = \sum_{ij}(M_b)_{ij}$$

Step 3:   Calculate Laplacian matrices (LM_w and LM_b) using the equations given below

$$LM_w = DM_w - M_w$$

$$LM_b = DM_b - M_b$$

Step 4:   Construct a matrix of k features by initially selecting randomly k features from original dataset (say R_k).

Step 5:   Declare an empty matrix (say N_k) to store top k features after finding scores of each feature

Step 6:   Repeat steps 6 to 10 until $R_k! = N_k$

Step 7:   Calculate $Y = XLM_bX^T$ and $Z = XLM_wX^T$.

Step 8:   Calculate Trace Ratios as $TR_y = TR(R_k^T Y R_k)$ and $TR_z = TR(R_k^T Z R_k)$.

Step 9:   Calculate $\beta = \frac{TR_y}{TR_z}$.

Step 10: Calculate Score of each feature as $F(f_i) = m_i^T(Y - \beta Z)m_i$.

Step 11: Select new top k features based on the score and store in N_k.

Step 12: Store final k features R_k for further processing

Step 13: Stop

ALGORITHM 5.1 Signal-to-noise ratio-trace ratio (SNR-TR) algorithm.

were performed for generation of average classification accuracy. The SNR itself is a well-defined technique, and when merged with the TR algorithm, it provided a better performance in less number of iterations as compared with the original TR algorithm.

Significance and Statistical Analysis of Signal-to-Noise Ratio

The SNR scoring method is used to redefine the TR algorithm where the weight values of each gene were considered. We used five datasets of different types for the above said purpose. Using the SNR, which itself is a statistical method, an effectiveness of the genes was measured for identifying a class out of another class. Hence, the adjacency matrix was redesigned within the class and between the class.

Implementation and Performance Analysis

Fig. 5.1 depicts the overall model of the proposed SNR-TR gene ranking algorithm. After normalizing the dataset using the min-max normalization, the refined dataset would be considered. The SNR-TR gene ranking phase involves certain subphase within it where using all scores of the SNR the weight matrix between the class and within the class is redefined. k number of genes are passed to the SNR-TR algorithm to obtain the rank of all the genes with respect to that k. The range of k in our

domain varies significantly as 50, 100, and 150. Based on this metric, the new ranked list is generated with less number of iterations. Later, this set of ranked list is passed to the variants of NN classifier for finding the accuracy and simultaneously validating the rank list. Ultimately, we found that instead of changing the base algorithm if we are able to suitably change the scoring pattern of the TR algorithm, then a huge difference in the performance can be obtained. Finally, it has been observed that the accuracy of the system improved a lot as compared with the existing TR algorithm, and the number of iterations for convergence of the algorithm is quite less. For most of the datasets, the accuracy obtained was 100%.

RESULT ANALYSIS

As discussed earlier, five datasets are considered for the purpose of evaluation where the results have been properly mentioned in the following tables along with the graphical representation of the same. The proposed SNR-TR gene ranking algorithm was passed to a different classification algorithm, and it was clearly visible that the result produced was quite good as compared with the original TR algorithm. Table 4.1 explains the detailed characteristics of the five datasets used.

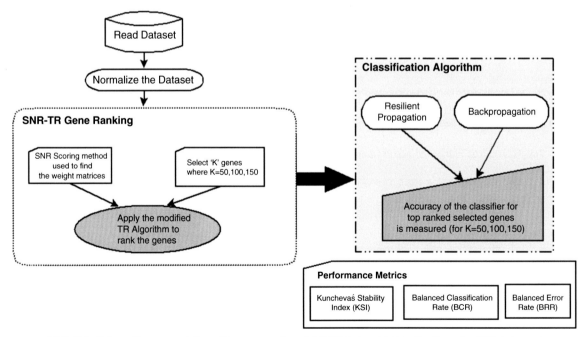

FIG. 5.1 Schematic representation of the proposed model. *BCR*, balanced classification rate; *BER*, balanced error rate; *KSI*, Kuncheva stability index; *SNR*, signal-to-noise ratio; *TR*, trace ratio.

TABLE 5.1
Performance Result of the Original TR Algorithm and SNR-TR Gene Ranking Algorithm for $k = 50$ Genes

	Resilient Propagation				Backpropagation			
	Original TR Algorithm		SNR-TR Gene Ranking Algorithm		Original TR Algorithm		SNR-TR Gene Ranking Algorithm	
Datasets	Acc	Itr	Acc	Itr	Acc	Itr	Acc	Itr
Colon cancer	98.36	122	**98.38**	**160**	79.03	2024	**98.38**	**544**
Leukemia	91.6	12	**95.8**	**34**	91.66	14	**100**	**42**
Medulloblastoma	99.10	44	**99.01**	**41**	76.47	725	**100**	**79**
Lymphoma	98.7	2335	**100**	**3242**	97.40	4901	**98.04**	**3982**
Prostate cancer	99.01	8197	**99.01**	**1731**	50.98	4987	**59.11**	**4775**

Acc, accuracy; *Itr*, iteration; *SNR*, signal-to-noise ratio; *TR*, trace ratio.

Tables 5.1—5.3 depict the performance of the original TR algorithm and the proposed SNR-TR algorithm for $k = 50$, 100, and 150 genes. Two classifiers have been habituated, and it was noticed that in most of the dataset, the accuracy is 100% in less number of iterations in comparison with the original TR algorithm. A clear comparison has been made where it can be distinguishably noticed that the SNR-TR gene ranking algorithm was performing better as compared with the original TR algorithm.

Table 5.4 shows the average accuracy obtained from five runs for $k = 200$ genes. Considerably, almost similar results were obtained as the previous table results, where the SNR-TR algorithm outperformed the TR algorithm with more accuracy in less number of iterations. We can also check the same for $k = 500$, and the results that would be produced out of this would also be more or nearly same as the earlier.

Figs. 5.2—5.6 show the accuracy graph of resilient propagation for the original TR algorithm and the SNR-TR method for $k = 50$, 100, and 150. It can be clearly seen that the proposed method provides a good accuracy measurement as compared with the other one.

Similarly, Figs. 5.7—5.11 also show the accuracy measurement for $k = 50$, 100, and 150 of backpropagation where again the SNR-TR algorithm outperformed the original TR algorithm. Other than that, it can also be observed that with increasing number of genes the accuracy is more or less same.

As stated earlier, all the figures depict the relationship between the original TR algorithm and the proposed SNR-TR gene ranking algorithm. For $k = 50$, 100, and 150 genes, the accuracy measurements are clearly seen where in most of the cases the accuracy is nearly same as the other gene.

TABLE 5.2
Performance Result of the Original TR Algorithm and SNR-TR Gene Ranking Algorithm for $k = 100$ Genes

	Resilient Propagation				Backpropagation			
	Original TR Algorithm		SNR-TR Gene Ranking Algorithm		Original TR Algorithm		SNR-TR Gene Ranking Algorithm	
Datasets	Acc	Itr	Acc	Itr	Acc	Itr	Acc	Itr
Colon cancer	98.3	108	**100**	**95**	97.24	1042	**100**	**215**
Leukemia	100	16	**100**	**12**	98.45	25	**100**	**27**
Medulloblastoma	99.01	41	**100**	**49**	98.24	3988	**100**	**41**
Lymphoma	98.70	1452	**100**	**1307**	99	4254	**99.25**	**2167**
Prostate cancer	99.01	4496	**99.01**	**1344**	95.25	2471	**97.11**	**2315**

Acc, accuracy; *Itr*, iteration; *SNR*, signal-to-noise ratio; *TR*, trace ratio.

TABLE 5.3
Performance Result of the Original TR Algorithm and SNR-TR Gene Ranking Algorithm for $k = 150$ Genes

| | Resilient Propagation | | | | Backpropagation | | | |
| | Original TR Algorithm | | SNR-TR Gene Ranking Algorithm | | Original TR Algorithm | | SNR-TR Gene Ranking Algorithm | |
Datasets	Acc	Itr	Acc	Itr	Acc	Itr	Acc	Itr
Colon cancer	99.01	123	**100**	104	97.58	1234	**98.38**	383
Leukemia	91.6	12	**93.75**	15	93.57	38	**100**	42
Medulloblastoma	99.01	56	**100**	38	98.65	4078	**100**	38
Lymphoma	98.7	2190	**100**	1515	93.47	4378	**95.15**	4289
Prostate cancer	99.01	4108	**99.01**	1026	88.25	2854	**93.51**	2711

Acc, accuracy; *Itr*, iteration; *SNR*, signal-to-noise ratio; *TR*, trace ratio.

TABLE 5.4
Average Performance Result of the Original TR Algorithm and SNR-TR Gene Ranking Algorithm for five Runs With $k = 200$ Genes

| | Resilient Propagation | | | | Backpropagation | | | |
| | Original TR Algorithm | | SNR-TR Gene Ranking Algorithm | | Original TR Algorithm | | SNR-TR Gene Ranking Algorithm | |
Datasets	Acc	Itr	Acc	Itr	Acc	Itr	Acc	Itr
Colon cancer	98.22	142	**100**	102	94.10	405	**97.44**	278
Leukemia	99.01	25	**99.58**	17	99.01	52	**100**	39
Medulloblastoma	98.14	56	**99.41**	37	93.22	79	**98.82**	55
Lymphoma	97.56	1254	**98.7**	988	93.11	3658	**95.39**	3179
Prostate cancer	99.01	1087	**99.80**	940	88.22	3567	**94.11**	2542

Acc, accuracy; *Itr*, iteration; *SNR*, signal-to-noise ratio; *TR*, trace ratio.

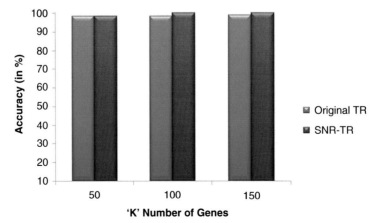

FIG. 5.2 Resilient propagation accuracy measurement of the original TR algorithm versus SNR-TR gene ranking algorithm of colon cancer dataset for $k = 50$, 100, and 150. *SNR*, signal-to-noise ratio; *TR*, trace ratio.

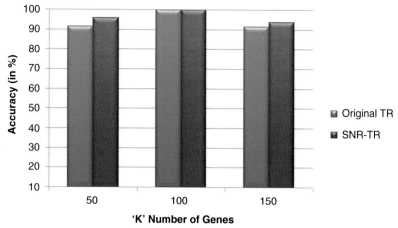

FIG. 5.3 Resilient propagation accuracy measurement of the original TR algorithm versus SNR-TR gene ranking algorithm of leukemia dataset for $k = 50$, 100, and 150. *SNR*, signal-to-noise ratio; *TR*, trace ratio.

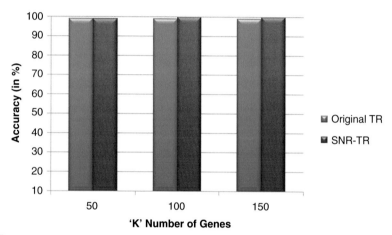

FIG. 5.4 Resilient propagation accuracy measurement of the original TR algorithm versus SNR-TR gene ranking algorithm of medulloblastoma dataset for $k = 50$, 100, and 150. *SNR*, signal-to-noise ratio; *TR*, trace ratio.

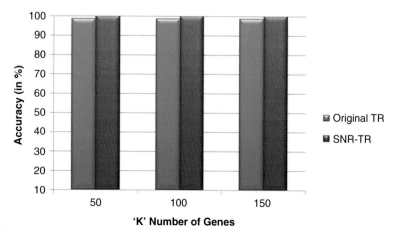

FIG. 5.5 Resilient propagation accuracy measurement of the original TR algorithm versus SNR-TR gene ranking algorithm of lymphoma dataset for $k = 50$, 100, and 150. *SNR*, signal-to-noise ratio; *TR*, trace ratio.

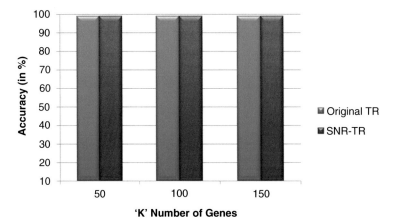

FIG. 5.6 Resilient propagation accuracy measurement of the original TR algorithm versus SNR-TR gene ranking algorithm of prostate cancer dataset for $k = 50, 100$, and 150. *SNR*, signal-to-noise ratio; *TR*, trace ratio.

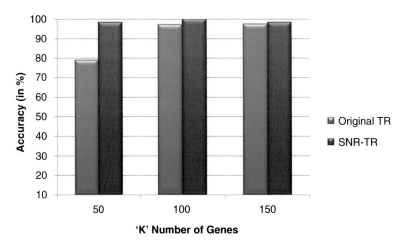

FIG. 5.7 Backpropagation accuracy measurement of the original TR algorithm versus SNR-TR gene ranking algorithm of colon cancer dataset for $k = 50, 100$, and 150. *SNR*, signal-to-noise ratio; *TR*, trace ratio.

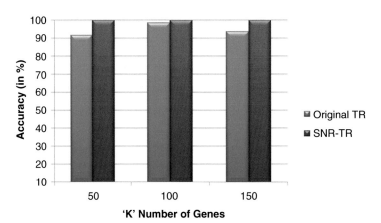

FIG. 5.8 Backpropagation accuracy measurement of the original TR algorithm versus SNR-TR gene ranking algorithm of leukemia dataset for $k = 50, 100$, and 150. *SNR*, signal-to-noise ratio; *TR*, trace ratio.

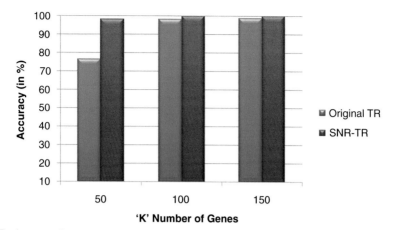

FIG. 5.9 Backpropagation accuracy measurement of the original TR algorithm versus SNR-TR gene ranking algorithm of medulloblastoma dataset for $k = 50$, 100, and 150. *SNR*, signal-to-noise ratio; *TR*, trace ratio.

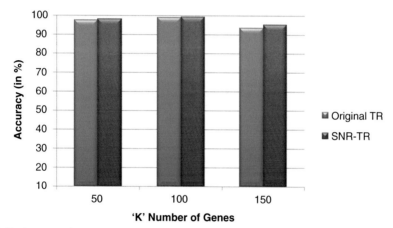

FIG. 5.10 Backpropagation accuracy measurement of original TR algorithm versus SNR-TR gene ranking algorithm of lymphoma dataset for $k = 50$, 100, and 150. *SNR*, signal-to-noise ratio; *TR*, trace ratio.

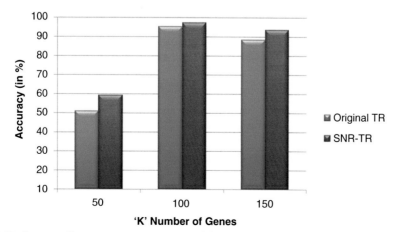

FIG. 5.11 Backpropagation accuracy measurement of the original TR algorithm versus SNR-TR gene ranking algorithm of prostate cancer dataset for $k = 50$, 100, and 150. *SNR*, signal-to-noise ratio; *TR*, trace ratio.

TABLE 5.5
Performance Assessment of Colon Cancer Dataset

k-Value	Index Metrics	Resilient Propagation		Backpropagation	
		Original TR	SNR-TR	Original TR	SNR-TR
50	KSI	0.26	0.69	0.19	0.68
	BCR	0.96	0.67	0.76	0.66
	BER	0.04	0.33	0.24	0.76
100	KSI	0.28	0.77	0.24	0.79
	BCR	0.93	0.82	0.93	0.83
	BER	0.07	0.18	0.07	0.17
150	KSI	0.25	0.77	0.22	0.69
	BCR	0.94	0.82	0.92	0.64
	BER	0.06	0.18	0.08	0.36

BCR, balanced classification rate; BER, balanced error rate; KSI, Kuncheva stability index; SNR, signal-to-noise ratio; TR, trace ratio.

We also performed validation of the algorithm using the three performance metrics such as KSI, BCR, and BER. Tables 5.5–5.9 depict the relationships that were extracted for the original TR algorithm and the SNR-TR algorithm. If carefully observed, in most of the cases for $k = 50$, 100, and 150 genes, it can be found that the result of SNR-TR algorithm for KSI and BCR approaches to 1 and for BER nears to 0. However, there are only certain cases in which the original TR algorithm is performing better for some gene values.

DISCUSSION
This chapter can be summarized as follows:
- For the original TR algorithm and SNR algorithm, the dataset has been normalized using the min-max normalization.
- From the original TR algorithm, by selecting the k number of genes (50, 100, and 150), the rank of the genes was calculated and was drawn to the classifier for obtaining the accuracy.

TABLE 5.6
Performance Assessment of Leukemia Dataset

k-Value	Index Metrics	Resilient Propagation		Backpropagation	
		Original TR	SNR-TR	Original TR	SNR-TR
50	KSI	0.30	0.64	0.25	0.73
	BCR	0.93	0.90	0.87	0.92
	BER	0.07	0.10	0.13	0.08
100	KSI	0.26	0.85	0.21	0.84
	BCR	0.97	0.95	0.93	0.93
	BER	0.03	0.05	0.07	0.07
150	KSI	0.22	0.85	0.15	0.84
	BCR	0.89	0.96	0.91	0.95
	BER	0.11	0.04	0.09	0.05

BCR, balanced classification rate; BER, balanced error rate; KSI, Kuncheva stability index; SNR, signal-to-noise ratio; TR, trace ratio.

TABLE 5.7
Performance Assessment of Medulloblastoma Dataset

k-Value	Index Metrics	Resilient Propagation		Backpropagation	
		Original TR	SNR-TR	Original TR	SNR-TR
50	KSI	0.22	**0.88**	0.22	**0.74**
	BCR	0.98	**0.95**	0.73	**0.86**
	BER	0.02	**0.05**	0.27	**0.14**
100	KSI	0.23	**0.86**	0.25	**0.67**
	BCR	0.94	**0.92**	0.95	**0.99**
	BER	0.06	**0.08**	0.05	**0.01**
150	KSI	0.27	**0.89**	0.21	**0.97**
	BCR	0.95	**0.94**	0.94	**0.99**
	BER	0.05	**0.06**	0.06	**0.01**

BCR, balanced classification rate; BER, balanced error rate; KSI, Kuncheva stability index; SNR, signal-to-noise ratio; TR, trace ratio.

TABLE 5.8
Performance Assessment of Lymphoma Dataset

k-Value	Index Metrics	Resilient Propagation		Backpropagation	
		Original TR	SNR-TR	Original TR	SNR-TR
50	KSI	0.29	**0.77**	0.18	**0.79**
	BCR	0.99	**0.90**	0.95	**0.94**
	BER	0.01	**0.10**	0.05	**0.06**
100	KSI	0.31	**0.81**	0.25	**0.85**
	BCR	0.94	**0.94**	0.97	**0.97**
	BER	0.06	**0.06**	0.03	**0.03**
150	KSI	0.27	**0.84**	0.27	**0.87**
	BCR	0.95	**0.99**	0.91	**0.99**
	BER	0.05	**0.01**	0.09	**0.01**

BCR, balanced classification rate; BER, balanced error rate; KSI, Kuncheva stability index; SNR, signal-to-noise ratio; TR, trace ratio.

- Now, we more or less modified the existing TR algorithm by using the powerful statistical measure called SNR.
- This measure was used to act as a new scoring or ranking criterion for the TR algorithm. Initially, Fisher score or Laplacian score was used as a scoring criterion that defines the weight matrix between the class and within the class.
- The SNR-TR algorithm ultimately generates the new rank list according to the specified k value, and this list was subsequently extended to the resilient propagation and backpropagation NN classifier, where the accuracy obtained for most of the dataset was 100%.
- In the end, the proposed method was validated and compared with the existing TR algorithm, where it was found that the proposed method outperformed the existing TR algorithm.

TABLE 5.9
Performance Assessment of Prostate Cancer Dataset

k-Value	Index Metrics	Resilient Propagation		Backpropagation	
		Original TR	SNR-TR	Original TR	SNR-TR
50	KSI	0.13	**0.64**	0.20	**0.72**
	BCR	0.95	**0.90**	0.48	**0.94**
	BER	0.05	**0.10**	0.52	**0.06**
100	KSI	0.20	**0.88**	0.20	**0.75**
	BCR	0.96	**0.97**	0.91	**0.95**
	BER	0.04	**0.03**	0.09	**0.05**
150	KSI	0.21	**0.84**	0.19	**0.87**
	BCR	0.97	**0.96**	0.84	**0.99**
	BER	0.03	**0.04**	0.16	**0.01**

BCR, balanced classification rate; *BER*, balanced error rate; *KSI*, Kuncheva stability index; *SNR*, signal-to-noise ratio; *TR*, trace ratio.

SUMMARY

A new gene selection method was proposed using the statistical SNR method and TR algorithm. The algorithm was used with five different datasets. The main objective of the technique was to generate an efficient rank list from a randomly selected set of genes. The rank list that was obtained was further validated using variants of NN classifiers, and the accuracy of the dataset was found to be mostly 100% with less number of iterations. We have also shown the average accuracy of the classifier for $k = 200$ genes for five runs. For rank generation, different k values are considered and it was concluded that by choosing a small set of k, we are able to acquire a better ranking pattern for the genes. In the next corresponding chapter, all the three techniques of gene selection (also derived from this chapter and Chapter 4) were used where the gene rank list generated out of them was used for the visualization of GRNs using Pearson correlation analysis. The GRNs for the five datasets were constructed, and visualization was made for the top 50, 100, and 150 genes selected from the three gene selection/ranking algorithms proposed.

CHAPTER 6

Visualization of Interactive Gene Regulatory Networks Using Gene Selection Techniques From Expression Data

SHRUTI MISHRA, PhD • SANDEEP KUMAR SATAPATHY, PhD • DEBAHUTI MISHRA, PhD

INTRODUCTION

Gene selection is one of the most fundamental methodologies for identifying the significant and relevant factors. Microarray data have ample number of genes available with few sample size information. It is quite known that all the factors present in the data matrix are nonrelevant and disease causing. Rather, there are few genes that highly contribute to the disease factor. Gene selection helps in identifying those genes whose disease contribution is clearly identifiable. The huge gene list contains various activated genes, and understanding their variations helps us to identify the functionalities of the cellular process.

The selected genes provide us with a great opportunity of researching and finding large scale of gene regulatory networks (GRNs) for identifying particular genes that are disease causing. This would assist researchers in finding the target genes, and understanding of the genetic interactions with other genes would further help us to consider various transcriptional factors needed for drug discovery operation. GRNs play an essential part in the designation of the molecular interaction and relationship among the genes based on some correlation factor. Merely, one of the central issues that always persists in GRNs is the modeling of the network.

GRNs are the systematic biologic interaction networks that describe the relationships between the genes in the form of a well-designed graph, where genes are represented by the nodes and the interactions or relationships among the genes are represented by the edges. GRNs help in placing the biologic and environmental effects of the interacting genes and likewise aid in identification of hub genes (genes that interact with many other factors). The hub genes are supposed to take on a major role in disease causing because of their key position in the GRNs.

As said earlier, modeling of GRNs is a major fear that still remains. Many computational approaches have come up for understanding the gene expression and finding their regulatory network and components. To develop a GRN, a reverse engineering strategy can be taken up where initially the genes would be selected from the huge available gene list and then those selected genes would be further used for generating the regulatory relationships. Based on this context, the different gene selection technique can be used for the selection of significant genes.

In this chapter, we have selected the three gene selection algorithms (as in Chapters 3−5) from which top 50, 100, and 150 genes were selected. Now, these selected genes were further used for modeling and designing GRNs from which interactive regulatory relationship can be extracted. This would further help in identifying the number of interactions taking place in the whole network, and also it would help us in identifying the hub genes. Those hub genes identified can further be biologically validated with the literature available. However, one of the major difficulties that aroused is that we were not able to biologically validate all the datasets that were used, as for some of the genes the names were not identifiable. In that case, we went for an assertion based on the available accession ids and the Affymetrix ids.

MATERIALS AND METHODS

This section deals with types of datasets used and the method or the gene selection algorithm that was used for extracting the genes. The same have been discussed in sections Support Vector Machine -Bayesian T-Test-Recursive Feature Elimination (SVM-BT-RFE) For Gene

Selection, Datasets Used, Proposed Methodologies of TR Algorithm for Gene Selection and Ranking, Datasets Used, Proposed SNR-TR Gene Ranking Algorithm and Datasets Used. These relevant sections deal with the algorithms and techniques along with the type of dataset that was needed for selection of genes. This section also introduces the basic concept that was needed for deriving the relationship between the selected genes. That concept is known as *Pearson correlation coefficient.*

Pearson Correlation Coefficient

It is one of the major statistical methods that measures the strength of the linear relationships between paired data. This technique can be used to evaluate and demonstrate the relationship between two quantitative continuous variables.[167,168] It can be further determined as the covariance between the two variables to the product of their standard deviations. As it takes the product-moment concept, it is as well known as Pearson product-moment correlation coefficient. The same has been defined in Eq. (6.1):

$$\rho_{X,Y} = \frac{COV(X,Y)}{\sigma_X \sigma_Y} \tag{6.1}$$

JUSTIFICATION OF THIS CHAPTER

The fundamental objective of this chapter is to visualize the network that has been constructed out of the selected genes. The selected genes are said to be a part of the disease-causing genes; hence, it becomes ultimately necessary to see or visualize which genes are heavily connected to the other genes in a large extent. Those genes that are heavily connected with other genes are usually called as *hub* genes. Other than that, visualization provides us a scope of identifying whether those genes are in some or other way not a part of the network. Hence, based on that we draw a conclusion that although they are disease-causing genes, they do not contribute much in causing the disease and their changes in environmental condition do not affect the structure and behavior of other genes.

EXPERIMENTATION

In this section we present the overall proposed model of the entire work that is carried over.

Proposed Model

Fig. 6.1 depicts the overall proposed model of our work. Initially, the datasets considered are normalized using min-max normalization process. Now this normalized set is passed through different ranking methods for the purpose ranking and to extract top 50 genes from the whole listed ranked set. Ultimately, for the purpose of visualization, a GRN using Pearson correlation coefficient is considered and the user-defined network is designed.

RESULT ANALYSIS

Here, we present and provide visualization of the biologic networks from the selected genes that were extracted using the three proposed gene selection algorithms, i.e., support vector machine-Bayesian *t*-test-recursive feature elimination (SVM-BT-RFE), canonical correlation analysis-trace ratio (CCA-TR), and signal-to-noise ratio-trace ratio (SNR-TR). Other than that, a brief and concrete statistical significance of the results is provided followed by the biologic visualization of the networks.

Statistical Significance of the Results

To find out the most significant genes, three types of gene selection algorithms were considered under different parametric conditions. They were further used to generate the regulatory relationships for which Pearson correlation coefficient was used with a threshold parameter of 0.6. In other words, genes having a range of 0.6 or above would be tagged as 1 and those having a range below 0.6 would be replaced by 0. This would help us to create an adjacency matrix through which interaction of genes with other genes can be effectively inferred. Cytoscape tool[169,170] has been used for the purpose of visualization and finding of the regulatory relationships from the user-defined network.

Biologic Visualization

Pearson correlation coefficient for generation of the regulatory interaction pattern was used, and the visualization was presented by Cytoscape using three gene selection algorithms for five types of datasets. Top 50, 100, and 150 ranked genes were selected, and their

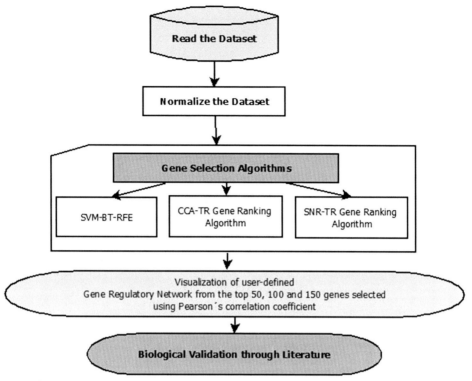

FIG. 6.1 Schematic model of the proposed work. *CCA-TR*, canonical correlation analysis-trace ratio; *SNR-TR*, signal-to-noise ratio-trace ratio; *SVM-BT-RFE*, support vector machine-Bayesian *t*-test-recursive feature elimination.

visualizations for each gene selection process are shown in the following sections (Figs. 6.2−6.46):

1. SVM-BT-RFE gene selection algorithm
   For 50 genes:
   For 100 genes:
   For 150 genes:
2. CCA-TR gene selection algorithm
   For 50 genes:
   For 100 genes:
   For 150 genes:
3. SNR-TR gene selection algorithm
   For 50 genes:
   For 100 genes:
   For 150 genes:

The networks constructed above have certain interesting facts such as there are certain genes that do not belong to the network structure. They appear to be aloof from the network layout. The fact is, although these genes contribute in classification accuracy prediction, they do not go to the segment where they are

more prone to disease causing. In other words, these genes are assumed to be less contributing toward disease.

DISCUSSION

The detailed number of interactions taking place throughout the network is evidenced in Table 6.1, and Table 6.6 represents the hub genes found along with their maximum interaction pathways or edges with the other corresponding genes. The same has been specifically mentioned for the three gene selection algorithms, SVM-BT-RFE, CCA-TR, and SNR-TR, with top 50, 100, and 150 ranked genes.

The interactions taking place between the top 50, 100, and 150 ranked genes can clearly be seen from Tables 6.1−6.3. This analysis further leads to find the hub genes associated with each network generated for the different algorithms (as shown in Tables 6.4−6.6). We can presumably think that these hub genes can be

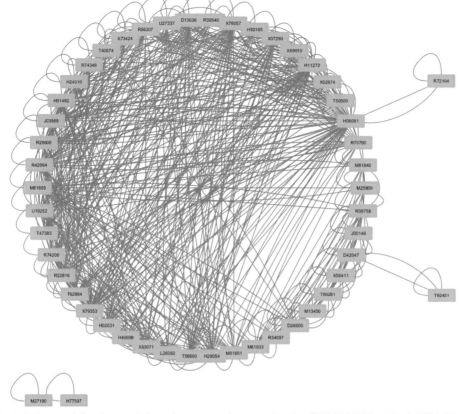

FIG. 6.2 A user-defined network for colon cancer dataset using the SVM-BT-RFE method. *SVM-BT-RFE*, support vector machine-Bayesian *t*-test-recursive feature elimination.

disease causing as predicted by the network construction. This chapter can be briefly summarized and highlighted as follows:

- The datasets were previously normalized using the min-max normalization process.
- The three proposed algorithms, i.e., SVM-BT-RFE, CCA-TR, and SNR-TR, were used to generate the top-ranked significant genes.
- The genes generated were further passed to the different performance metrics, and their performance was assessed.
- The top 50 ranked genes were considered for discovery of the user-defined regulatory relationships using Pearson correlation coefficient.
- A threshold limit of 0.6 was used in Pearson correlation, i.e., genes having 0.6 or above are replaced by 1 and those having below 0.6 by 0.

- Based on this, the adjacency matrix was generated, and using the same, a Cytoscape tool was used to visualize the network.
- Lastly, the hub genes were extracted from the user-defined network, which can be in the future be biologically validated using official gene symbols and functional annotation tools. We have only stated the corresponding accession ids and Affymetrix ids for the concerned hub genes.

SUMMARY

In this chapter, we provided an overview of the previously proposed gene selection algorithm from which top 50, 100, and 150 ranked genes were used for the current analysis. Using the genes, the regulatory relationships were depicted, and out of these relationships,

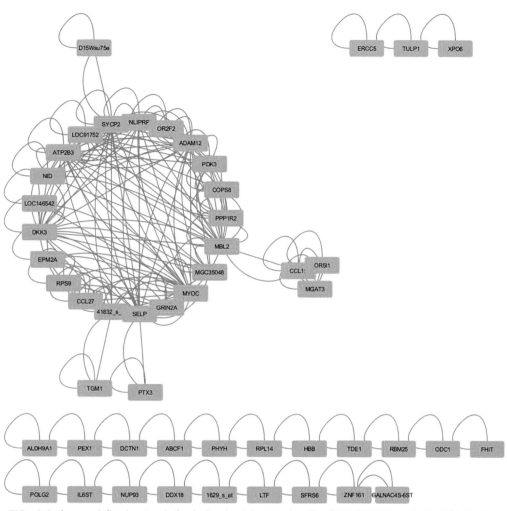

FIG. 6.3 A user-defined network for leukemia dataset using the SVM-BT-RFE method. *SVM-BT-RFE*, support vector machine-Bayesian *t*-test-recursive feature elimination.

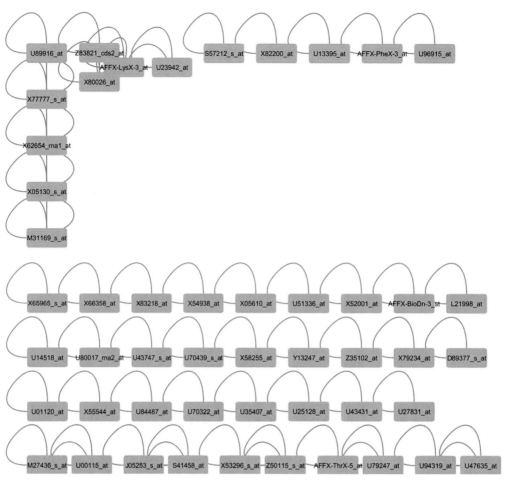

FIG. 6.4 A user-defined network for medulloblastoma dataset using the SVM-BT-RFE method. *SVM-BT-RFE*, support vector machine-Bayesian *t*-test-recursive feature elimination.

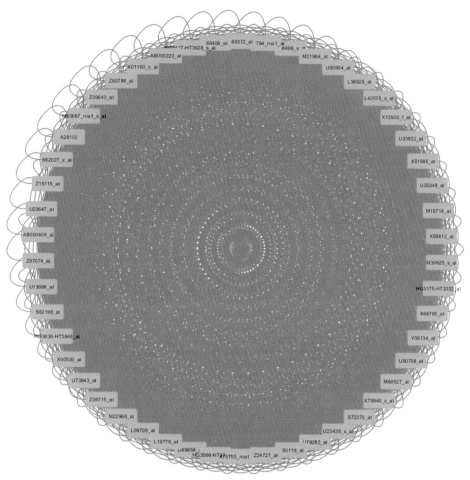

FIG. 6.5 A user-defined network for lymphoma dataset using the SVM-BT-RFE method. *SVM-BT-RFE*, support vector machine-Bayesian *t*-test-recursive feature elimination.

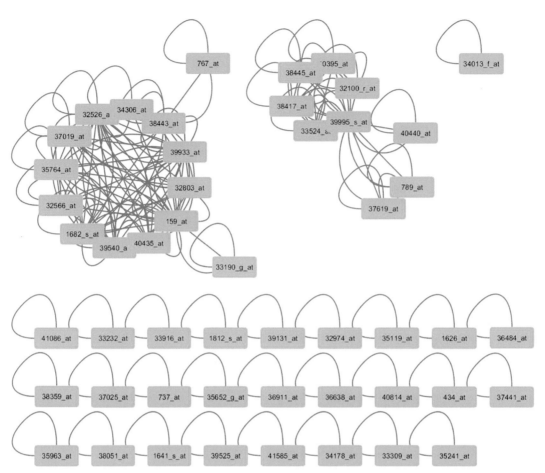

FIG. 6.6 A user-defined network for prostate cancer dataset using the SVM-BT-RFE method. *SVM-BT-RFE*, support vector machine-Bayesian *t*-test-recursive feature elimination.

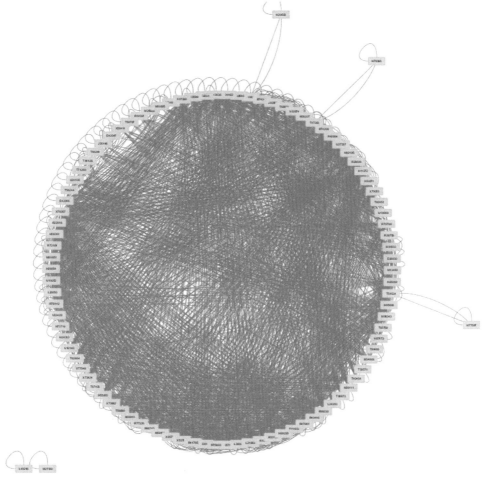

FIG. 6.7 A user-defined network for colon cancer dataset using the SVM-BT-RFE method. *SVM-BT-RFE*, support vector machine-Bayesian *t*-test-recursive feature elimination.

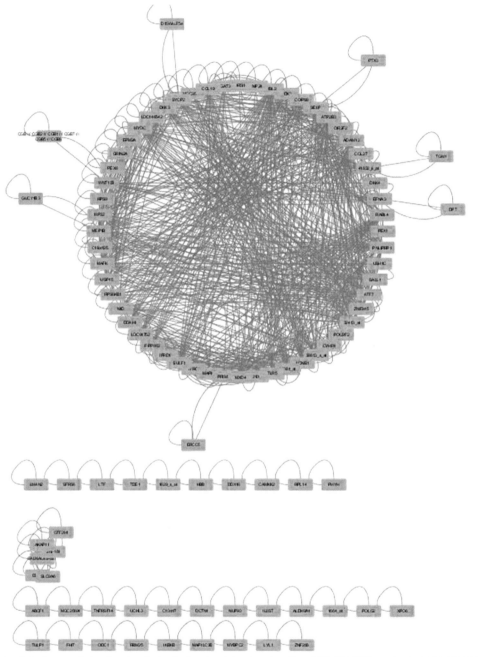

FIG. 6.8 A user-defined network for leukemia dataset using the SVM-BT-RFE method. *SVM-BT-RFE*, support vector machine-Bayesian *t*-test-recursive feature elimination.

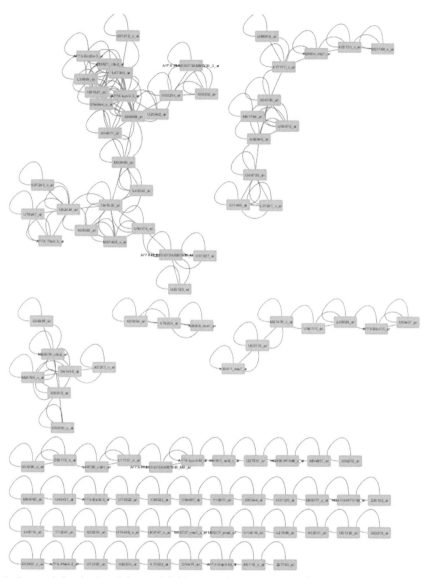

FIG. 6.9 A user-defined network for medulloblastoma dataset using the SVM-BT-RFE method. *SVM-BT-RFE*, support vector machine-Bayesian *t*-test-recursive feature elimination.

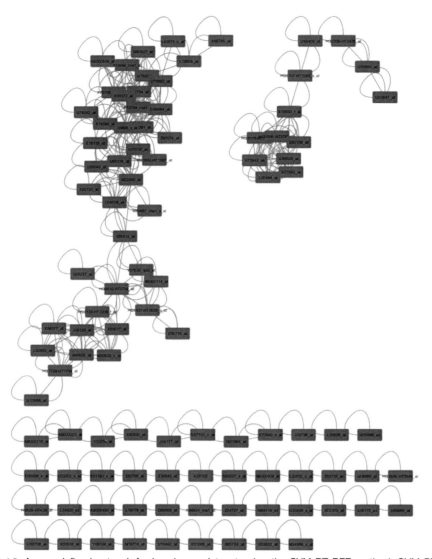

FIG. 6.10 A user-defined network for lymphoma dataset using the SVM-BT-RFE method. *SVM-BT-RFE*, support vector machine-Bayesian *t*-test-recursive feature elimination.

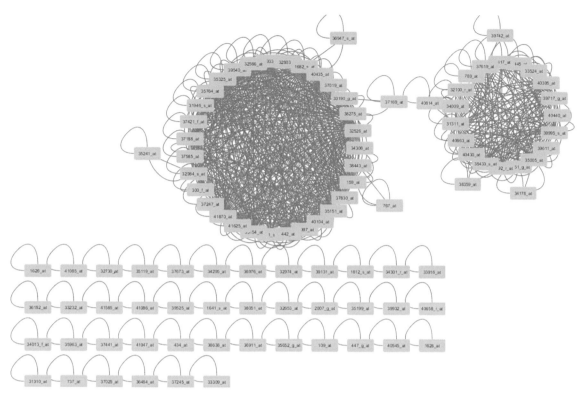

FIG. 6.11 A user-defined network for prostate cancer dataset using the SVM-BT-RFE method. *SVM-BT-RFE*, support vector machine-Bayesian *t*-test-recursive feature elimination.

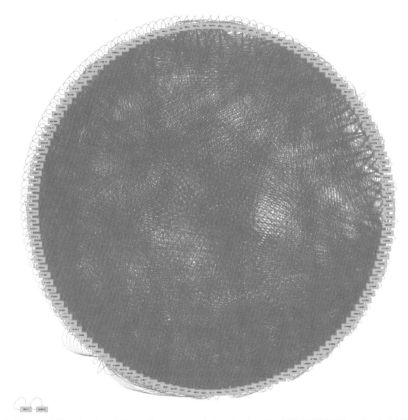

FIG. 6.12 A user-defined network for colon cancer dataset using the SVM-BT-RFE method. *SVM-BT-RFE*, support vector machine-Bayesian *t*-test-recursive feature elimination.

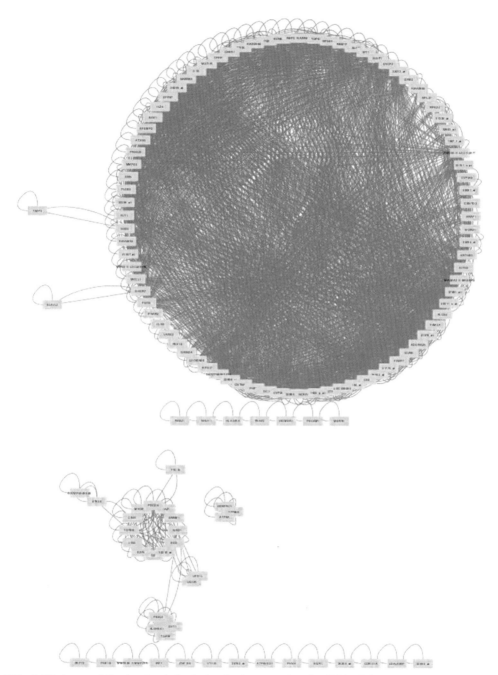

FIG. 6.13 A user-defined network for leukemia dataset using the SVM-BT-RFE method. *SVM-BT-RFE*, support vector machine-Bayesian *t*-test-recursive feature elimination.

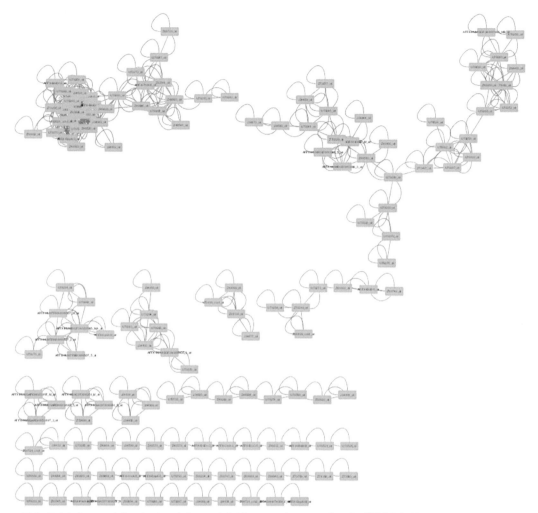

FIG. 6.14 A user-defined network for medulloblastoma dataset using the SVM-BT-RFE method. *SVM-BT-RFE*, support vector machine-Bayesian *t*-test-recursive feature elimination.

FIG. 6.15 A user-defined network for lymphoma dataset using the SVM-BT-RFE method. *SVM-BT-RFE*, support vector machine-Bayesian *t*-test-recursive feature elimination.

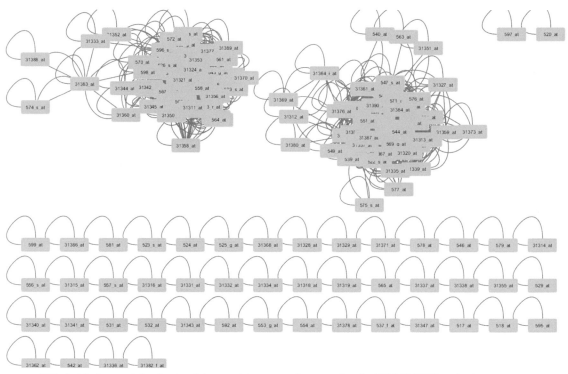

FIG. 6.16 A user-defined network for prostate cancer dataset using the SVM-BT-RFE method. *SVM-BT-RFE*, support vector machine-Bayesian *t*-test-recursive feature elimination.

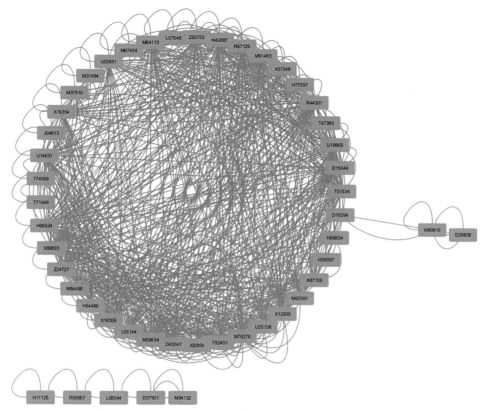

FIG. 6.17 A user-defined network for colon cancer dataset using the CCA-TR gene ranking method. *CCA-TR*, canonical correlation analysis-trace ratio.

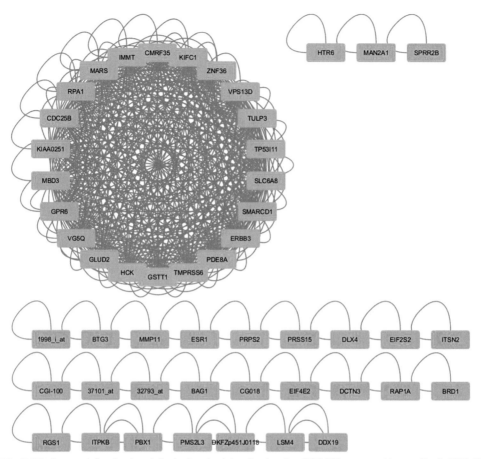

FIG. 6.18 A user-defined network for leukemia dataset using the CCA-TR gene ranking method. *CCA-TR*, canonical correlation analysis-trace ratio.

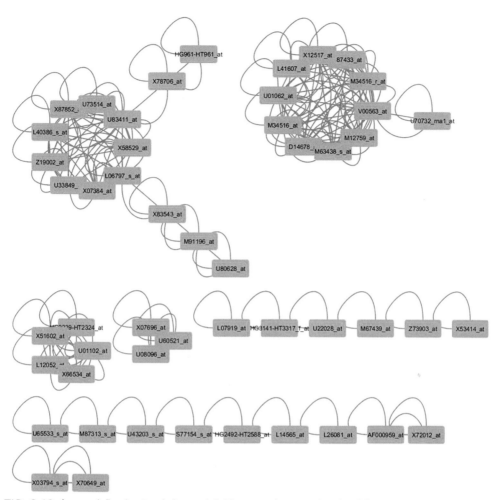

FIG. 6.19 A user-defined network for medulloblastoma dataset using the CCA-TR gene ranking method. *CCA-TR*, canonical correlation analysis-trace ratio.

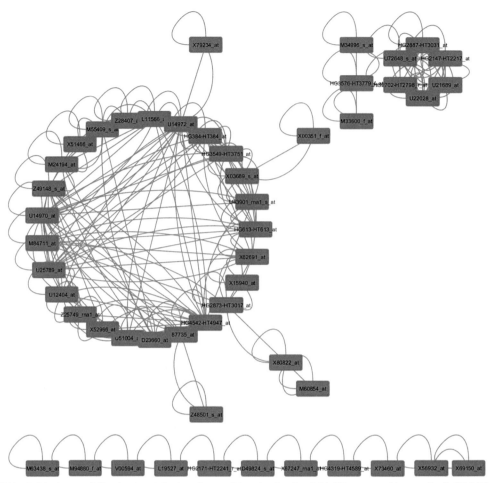

FIG. 6.20 A user-defined network for lymphoma dataset using the CCA-TR gene ranking method. *CCA-TR*, canonical correlation analysis-trace ratio.

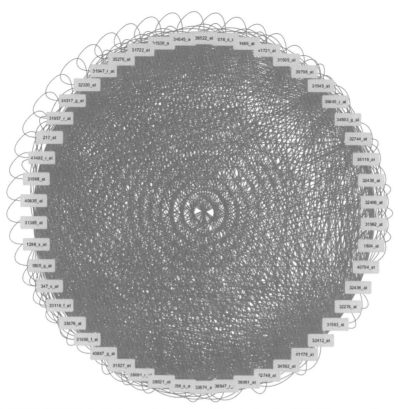

FIG. 6.21 A user-defined network for prostate cancer dataset using the CCA-TR gene ranking method. *CCA-TR*, canonical correlation analysis-trace ratio.

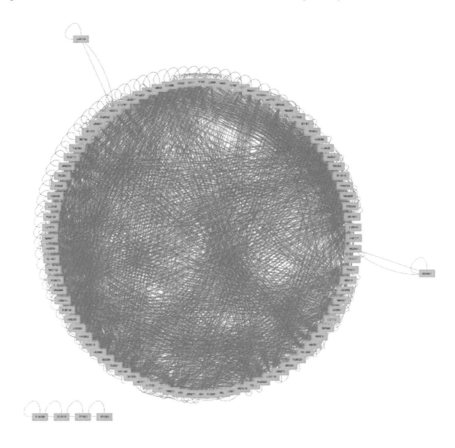

FIG. 6.22 A user-defined network for colon cancer dataset using the CCA-TR gene ranking method. *CCA-TR*, canonical correlation analysis-trace ratio.

FIG. 6.23 A user-defined network for leukemia dataset using the CCA-TR gene ranking method. *CCA-TR*, canonical correlation analysis-trace ratio.

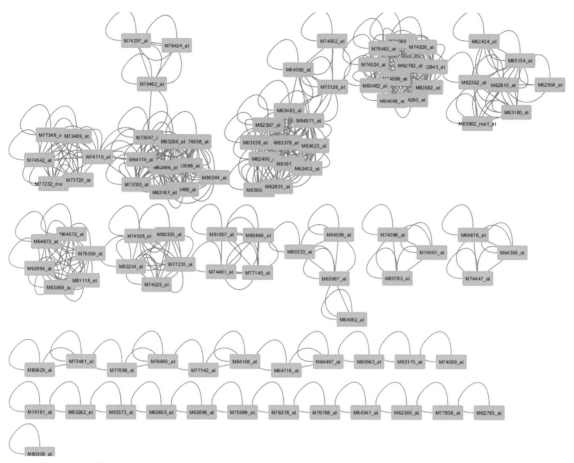

FIG. 6.24 A user-defined network for medulloblastoma dataset using the CCA-TR gene ranking method. *CCA-TR*, canonical correlation analysis-trace ratio.

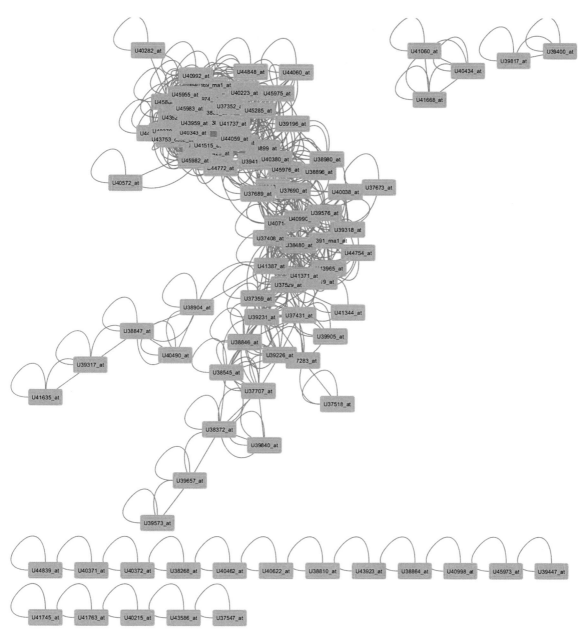

FIG. 6.25 A user-defined network for lymphoma dataset using the CCA-TR gene ranking method. *CCA-TR*, canonical correlation analysis-trace ratio.

FIG. 6.26 A user-defined network for prostate cancer dataset using the CCA-TR gene ranking method. *CCA-TR*, canonical correlation analysis-trace ratio.

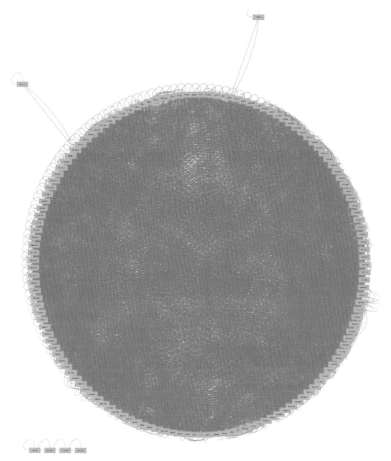

FIG. 6.27 A user-defined network for colon cancer dataset using the CCA-TR gene ranking method. *CCA-TR*, canonical correlation analysis-trace ratio.

FIG. 6.28 A user-defined network for leukemia dataset using the CCA-TR gene ranking method. *CCA-TR*, canonical correlation analysis-trace ratio.

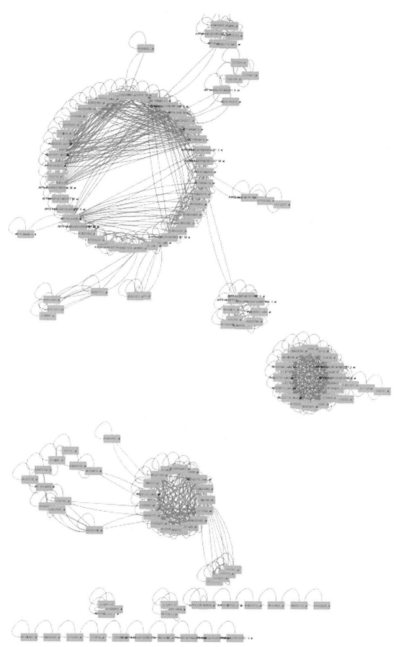

FIG. 6.29 A user-defined network for medulloblastoma dataset using the CCA-TR gene ranking method. *CCA-TR*, canonical correlation analysis-trace ratio.

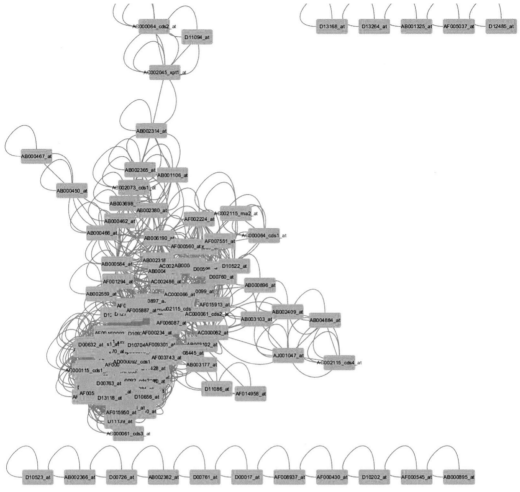

FIG. 6.30 A user-defined network for lymphoma dataset using the CCA-TR gene ranking method. *CCA-TR*, canonical correlation analysis-trace ratio.

FIG. 6.31 A user-defined network for prostate cancer dataset using the CCA-TR gene ranking method. *CCA-TR*, canonical correlation analysis-trace ratio.

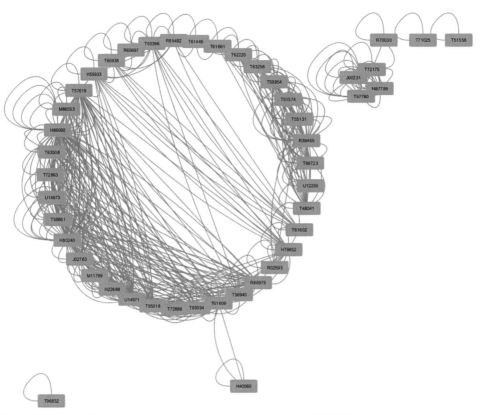

FIG. 6.32 A user-defined network for colon cancer dataset using the SNR-TR gene ranking method. *SNR-TR*, signal-to-noise ratio-trace ratio.

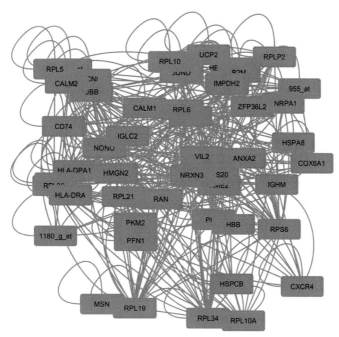

FIG. 6.33 A user-defined network for leukemia dataset using the SNR-TR gene ranking method. *SNR-TR*, signal-to-noise ratio-trace ratio.

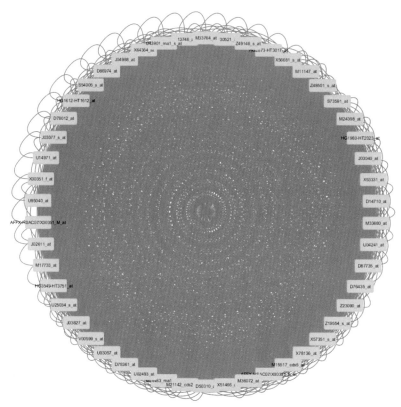

FIG. 6.34 A user-defined network for medulloblastoma dataset using the SNR-TR gene ranking method. *SNR-TR*, signal-to-noise ratio-trace ratio.

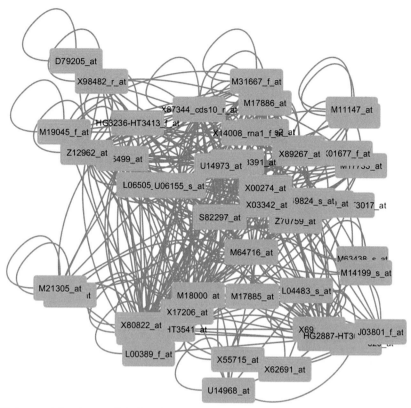

FIG. 6.35 A user-defined network for lymphoma dataset using the SNR-TR gene ranking method. *SNR-TR*, signal-to-noise ratio-trace ratio.

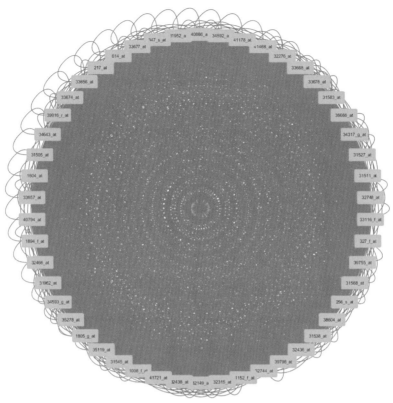

FIG. 6.36 A user-defined network for prostate cancer dataset using the SNR-TR gene ranking method. *SNR-TR*, signal-to-noise ratio-trace ratio.

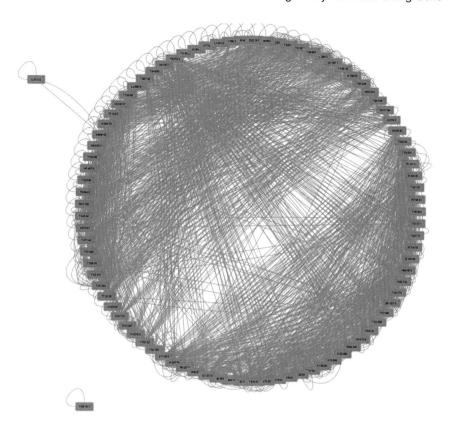

FIG. 6.37 A user-defined network for colon cancer dataset using the SNR-TR gene ranking method. *SNR-TR*, signal-to-noise ratio-trace ratio.

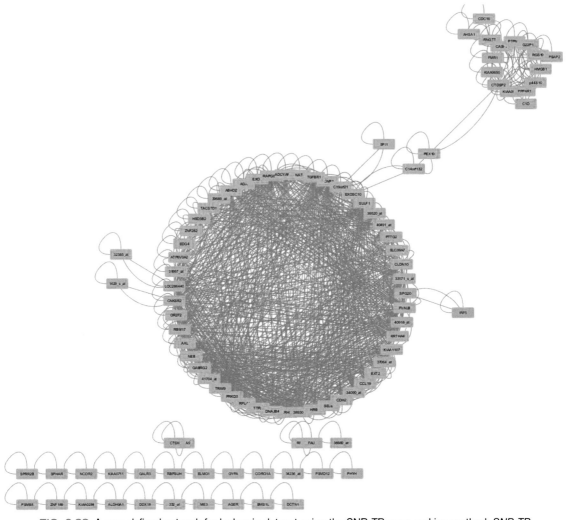

FIG. 6.38 A user-defined network for leukemia dataset using the SNR-TR gene ranking method. *SNR-TR*, signal-to-noise ratio-trace ratio.

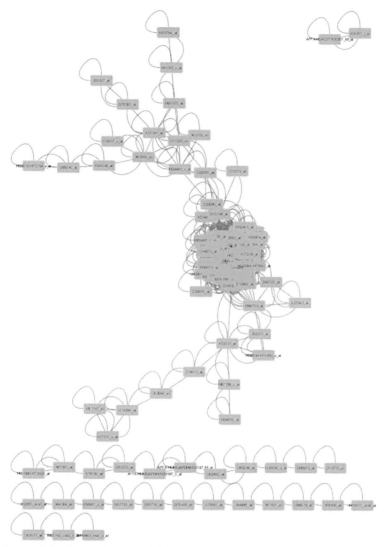

FIG. 6.39 A user-defined network for medulloblastoma dataset using the SNR-TR gene ranking method. *SNR-TR*, signal-to-noise ratio-trace ratio.

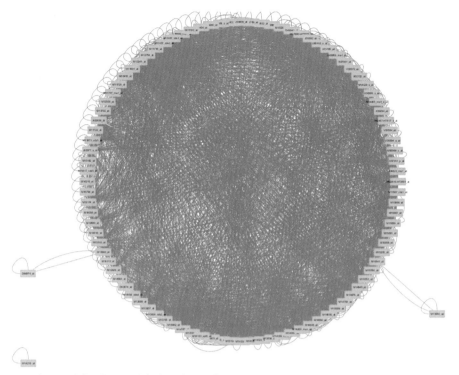

FIG. 6.40 A user-defined network for lymphoma dataset using the SNR-TR gene ranking method. *SNR-TR*, signal-to-noise ratio-trace ratio.

FIG. 6.41 A user-defined network for prostate cancer dataset using the SNR-TR gene ranking method. *SNR-TR*, signal-to-noise ratio-trace ratio.

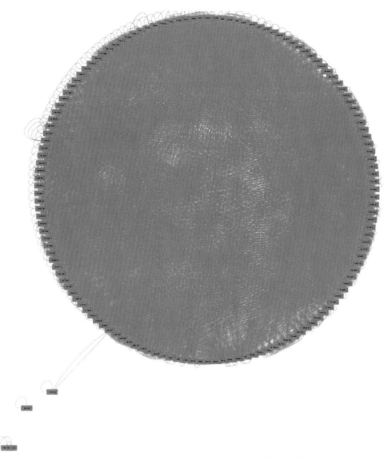

FIG. 6.42 A user-defined network for colon cancer dataset using the SNR-TR gene ranking method. *SNR-TR*, signal-to-noise ratio-trace ratio.

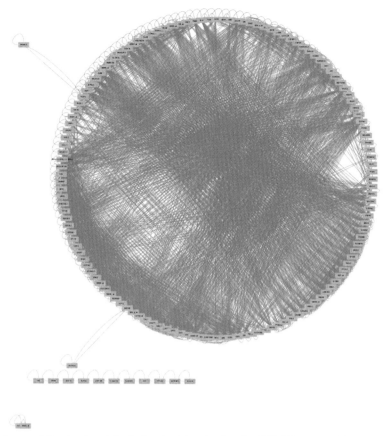

FIG. 6.43 A user-defined network for leukemia dataset using the SNR-TR gene ranking method. *SNR-TR*, signal-to-noise ratio-trace ratio.

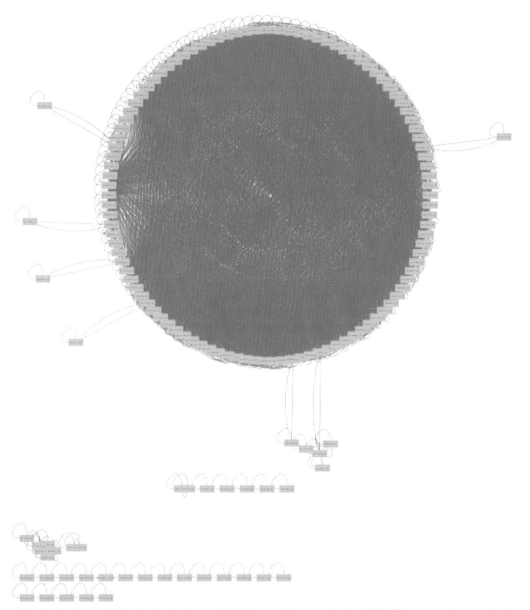

FIG. 6.44 A user-defined network for medulloblastoma dataset using the SNR-TR gene ranking method. *SNR-TR*, signal-to-noise ratio-trace ratio.

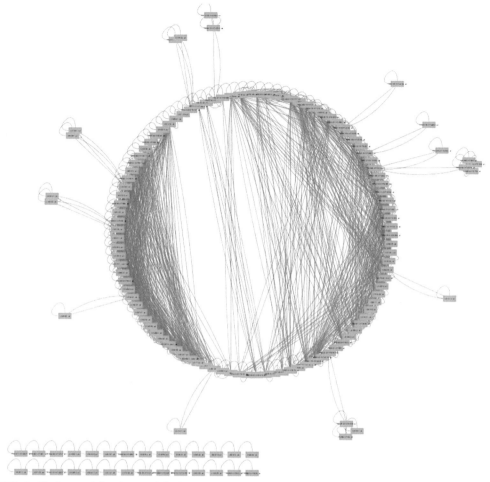

FIG. 6.45 A user-defined network for lymphoma dataset using the SNR-TR gene ranking method. *SNR-TR*, signal-to-noise ratio-trace ratio.

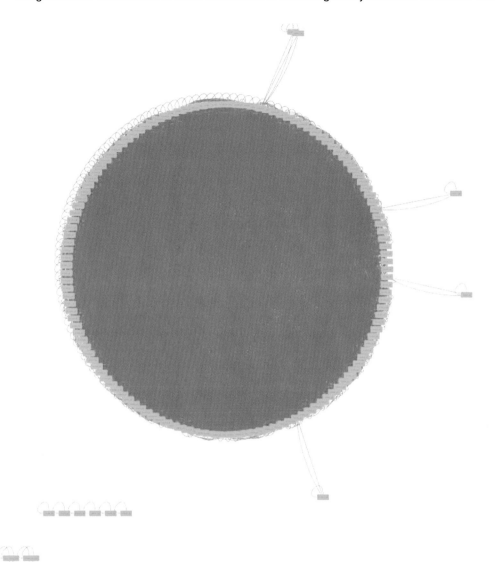

FIG. 6.46 A user-defined network for prostate cancer dataset using the SNR-TR gene ranking method. *SNR-TR*, signal-to-noise ratio-trace ratio.

TABLE 6.1

The Number of Edges or Interaction of Five Datasets Using the Three Basic Gene Selection Algorithms for Top 50 Ranked Genes

Datasets	Number of Edges or Interaction With the Gene Selection Algorithms		
	SVM-BT-RFE	CCA-TR	SNR-TR
Colon cancer	606	754	446
Leukemia	198	512	330
Medulloblastoma	76	220	2500
Lymphoma	2500	230	322
Prostate cancer	172	1560	2500

CCA-TR, canonical correlation analysis-trace ratio; *SNR-TR*, signal-to-noise ratio-trace ratio; *SVM-BT-RFE*, support vector machine-Bayesian *t*-test-recursive feature elimination.

TABLE 6.3

The Number of Edges or Interaction of Five Datasets Using the Three Basic Gene Selection Algorithms for Top 150 Ranked Genes

Datasets	Number of Edges or Interaction With the Gene Selection Algorithms		
	SVM-BT-RFE	CCA-TR	SNR-TR
Colon cancer	5,092	5,430	5,976
Leukemia	2,360	22,202	3,116
Medulloblastoma	574	1,050	4,864
Lymphoma	660	3,156	1,404
Prostate cancer	2,148	11,974	8,536

CCA-TR, canonical correlation analysis-trace ratio; *SNR-TR*, signal-to-noise ratio-trace ratio; *SVM-BT-RFE*, support vector machine-Bayesian *t*-test-recursive feature elimination.

TABLE 6.2

The Number of Edges or Interaction of Five Datasets Using the Three Basic Gene Selection Algorithms for Top 100 Ranked Genes

Datasets	Number of Edges or Interaction With the Gene Selection Algorithms		
	SVM-BT-RFE	CCA-TR	SNR-TR
Colon cancer	2,141	2,158	1,942
Leukemia	846	10,000	992
Medulloblastoma	248	562	1,390
Lymphoma	368	1,062	2,696
Prostate cancer	842	5,548	3,998

CCA-TR, canonical correlation analysis-trace ratio; *SNR-TR*, signal-to-noise ratio-trace ratio; *SVM-BT-RFE*, support vector machine-Bayesian *t*-test-recursive feature elimination.

TABLE 6.4
Hub Genes Found in Each of the Datasets for the Three Gene Selection Algorithms for Top 50 Ranked Genes

| | Hub Genes Found From the Three Gene Selection Algorithms | | | | | |
| | SVM-BT-RFE | | CCA-TR | | SNR-TR | |
Datasets	Accession Ids/ Affymetrix Ids	Number of Interactions	Accession Ids/ Affymetrix Ids	Number of Interactions	Accession Ids/ Affymetrix Ids	Number of Interactions
Colon cancer	H11272	27	R44301	30	U14971	20
	R56207	24	M91463	29	T57619	19
Leukemia	MYOC	14	ZNF36	22	RPS6	14
	MBL2	12	MBD3	22	RPL19	14
	ATP2B3	12	GPR6	22	RPL34	13
Medulloblastoma	X80026_at	3	V00563_at	11	X00351_f	50
	X62654_rna1_at	3	M34516_at	10	J02611_at	50
Lymphoma	U20647_at	50	U14972_at	12	Z12962_at	17
	297074_at	50	U14970_at	11	U14973_at	16
Prostate cancer	159_at	13	31583_at	44	1804_at	50
	32526_at	12	31545_at	44	40794_at	50

CCA-TR, canonical correlation analysis-trace ratio; *SNR-TR*, signal-to-noise ratio-trace ratio; *SVM-BT-RFE*, support vector machine-Bayesian *t*-test-recursive feature elimination.

TABLE 6.5
Hub Genes Found in Each of the Datasets for the Three Gene Selection Algorithms for Top 100 Ranked Genes

| | Hub Genes Found From the Three Gene Selection Algorithms | | | | | |
| | SVM-BT-RFE | | CCA-TR | | SNR-TR | |
Datasets	Accession Ids/ Affymetrix Ids	Number of Interactions	Accession Ids/Affymetrix Ids	Number of Interactions	Accession Ids/ Affymetrix Ids	Number of Interactions
Colon cancer	H06061	54	T92259	46	H46732	37
	H11272	50	H51196	46	D14696	35
Leukemia	ADAM12	28	SCFD1	100	38930_at	33
	MYOC	27	NROB2	100	PVALB	31
Medulloblastoma	L47345	8	M61916_at	11	Z49148_s_at	39
	X96969_at	8	M62400_at	11	M24194_at	39
Lymphoma	H03805_s_at	16	U44059_at	31	X67951_at	60
	X75755_rna1_at	11	U41737_at	30	J04823_rna1_at	59
Prostate cancer	1687_s_at	30	38940_at	82	39226_at	58
	32526_at	30	38962_at	81	39208_i_at	58

CCA-TR, canonical correlation analysis-trace ratio; *SNR-TR*, signal-to-noise ratio-trace ratio; *SVM-BT-RFE*, support vector machine-Bayesian *t*-test-recursive feature elimination.

TABLE 6.6
Hub Genes Found in Each of the Datasets for the Three Gene Selection Algorithms for Top 150 Ranked Genes

	Hub Genes Found From the Three Gene Selection Algorithms						
	SVM-BT-RFE		**CCA-TR**		**SNR-TR**	
Datasets	**Accession Ids/ Affymetrix Ids**	**Number of Interactions**	**Accession Ids/ Affymetrix Ids**	**Number of Interactions**	**Accession Ids/ Affymetrix Ids**	**Number of Interactions**
Colon cancer	H39830	71	X66503	83	X03674	86
	X78817	70	H40560	83	X62153	80
Leukemia	32896_at	51	DYT1	149	UPF2	48
	RAPGEF4	50	CORO1	149	ATP9A	48
Medulloblastoma	Z83806_at	17	M33197_3_at	18	D14695_at	82
	Z83800_at	15	M10098_3_at	16	D14043_at	81
Lymphoma	U49250_at	19	AF000177_at	56	HG4102_HT4372_at	28
	U48936_at	17	AF009368_at	58	HG4114_HT4384_at	28
Prostate cancer	31353_f_at	42	39291_at	117	31370_at	101
	31375_at	41	39294_at	118	31334_at	100

CCA-TR, canonical correlation analysis-trace ratio; *SNR-TR*, signal-to-noise ratio-trace ratio; *SVM-BT-RFE*, support vector machine-Bayesian *t*-test-recursive feature elimination.

proper insight details were extracted along with the visualization of the network. Hub genes were extracted out of the current networks that were having a maximum number of interactions with other genes. These hub genes are extracted from the user-defined network and hence can be biologically validated in the future using the official gene symbols and the functional annotation tools available. In Chapter 7, a detailed conclusion and future scope toward the work is stated.

CHAPTER 7

Conclusion and Future Work

SHRUTI MISHRA, PhD

CONCLUSION

As discussed earlier, *Data mining* and *Bioinformatics* have played a critical and central role in the evolution of systems biology. Maybe this was the reason why the focus shifted toward the microarray technology and gene expression data. Understanding the details about the gene expression data is one of the most challenging concepts in systems biology and inferring conclusion out of them is another hurdle.

Throughout this research subject area, a focus was made toward the understanding of the details around the disease-causing cancerous genes. As we know that the curse of dimensionality always prevails upon the gene expression data, where the number of genes is huge and the expression levels in the tally of thousands but there is a poor availability of the samples. Hence, to satisfy this aim, a proper outlook and algorithm with specific logic was called for. Thus, the target changed to determining an appropriate method for the uncovering of the genes. It was called as a gene selection algorithm.

In this study, three fundamental gene selection algorithms were proposed based on biologic and statistical aspects of gene expression data. Support vector machine-Bayesian *t*-test-recursive feature elimination (SVM-BT-RFE), canonical correlation analysis-trace ratio (CCA-TR), and signal-to-noise ratio-trace ratio (SNR-TR) gene selection algorithms were specifically applied. The SVM-BT-RFE was based on a famous framework of Bayesian *t*-test and SVM-RFE where selection of the gene was established based on the iterative elimination process and the significant topmost gene values. The selected genes extracted from the same were tried out using the SVM classifier for finding the predictive accuracy. It was noted that the effects produced were far better than the generalized SVM-RFE. Only a major drawback of this algorithm was it heavily relied on the SVM weight vectors; as a result, it imposed a great dependency factor. So, our focus shifted toward another famous algorithm called the TR algorithm where the selection of the genes was based on the trace value obtained. The traditional TR algorithm used Fisher score or Laplacian score as the major ranking criterion for finding the top-ranked genes. Instead of applying the existing scoring method, if we could focus on some different scoring method, then it could be more beneficial for enhancing the classifier's performance level and accuracy. Hence, the CCA was used instead of Fisher score or Laplacian score. The results produced out of the same also outperformed the existing results of the generalized TR algorithm. Again, the CCA had its own share of demerits. It involved a lot of hefty calculation such as covariance and variance estimation. As a consequence, the SNR was proposed. The SNR is itself a famous statistical technique that took a mere computation of mean and variance for finding the rank of the genes. The same was used and merged with the existing TR algorithm for generating a new set of rank list. It was specifically noted that from the results produced the proposed algorithm outperformed the existing TR algorithm in terms of classification accuracy. The CCA-TR and SNR-TR were further examined with some probability measures such as Kuncheva stability index (KSI), balanced classification rate (BCR), and balanced error rate (BER) for further reassurance and revalidating the classifier's accuracy. Lastly, a thought of visualizing the selected genes came across. Hence, for that we used the famous gene regulatory network concept. Pearson correlation coefficient was used for depicting the relationship among the selected genes, and then the relationship list was passed to a software tool called Cytoscape for visualization purpose. After the construction of the network, a clear picture can be drawn out as which are the genes that are connected heavily in the network and which are the genes that are not at all connected.

FUTURE WORK

This study can be further extended to a large domain where proper biological analysis and validation can be made for all the selected genes. Other than that, based on the biologic analysis, relationships and condition can be inferred, using which genes affect the other genes in the network. A proper detailed study can be made out of the visualized network with a thorough investigation about the genes' connection and their aspects of affecting each other in different environmental conditions. Other than this aspect, pathways can be established and domains can be detected for the genes responsible for the disease.

References

1. Chandra B, Gupta M. An efficient statistical feature selection approach for classification of gene expression data. *J Biomed Inf.* 2011;44:529–535.
2. Kar S, Das Sharma K, Maitra M. Gene selection from microarray gene expression data for classification of cancer sub-groups employing PSO and adaptive K-nearest neighbourhood technique. *Expert Syst Appl.* 2015;42:612–627.
3. Du D, Li K, Li X, Fei M. A novel forward gene selection algorithm for microarray data. *Neurocomputing.* 2014; 133:446–458.
4. Kamkar I, Gupta SK, Phung D, Venkatesh S. Stable feature selection for clinical prediction: exploiting ICD tree structure using Tree-Lasso. *J Biomed Inf.* 2014;53:277–290.
5. Schena M, Shalon D, Davis R, Brown P. Quantitative monitoring of gene expression patterns with a complementary DNA microarray. *Science.* 1995;270(5235):467–470.
6. Bammler T, Beyer RP, Kerr X, et al. Standardizing global gene expression analysis between laboratories and across platforms. *Nat Methods.* 2005;2(5):351–356.
7. Pease AC, Solas D, Sullivan EJ, Cronin MT, Holmes CP, Fodor SP. Light-generated oligonucleotide arrays for rapid DNA sequence analysis. *PNAS.* 1994;91(11):5022–5026.
8. Dan S, Tsunoda T, Kitahara O, Yanagawa R, Zembutsu H, Katagiri T. An integrated database of chemo-sensitivity to 55 anticancer drugs and gene expression profiles of 39 human cancer cell lines. *Cancer Res.* 2002;62(4):1139–1147.
9. De-Risi J, Penland L, Brown P, Bittner M, Meltzer P, Ray M. Use of a CDNA microarray to analyse gene expression patterns in human cancer. *Nat Genet.* 1996;14(4):457–460.
10. Goldmann T, Gonzalez JS. DNA-printing: utilization of a standard inkjet printer for the transfer of nucleic acids to solid supports. *J Biochem Biophysical Methods.* 2000;42(3): 105–1016.
11. Mukherjee A, Vasquez KM. Triplex technology in studies of DNA damage, DNA repair, and mutagenesis. *Biochimie.* 2011;93(8):1197–1208.
12. Shalon D, Smith SJ, Brown PO. A DNA microarray system for analyzing complex DNA samples using two-colour fluorescent probe hybridization. *Genome Res.* 1996;6(7): 639–645.
13. Churchill GA. Fundamentals of experimental design for cDNA microarrays. *Nat Genet.* 2002;32:490–495.
14. Slack JMW. *Genes-A Very Short Introduction.* 2nd ed. Oxford University Press; 2014.
15. Gericke NM, Hagberg M. Definition of historical models of gene function and their relation to students' understanding of genetics. *Science & Education.* 2006;16(7–8): 849–881.
16. Pearson H. Genetics: what is a gene? *Nature.* 2006; 441(7092):398–401.
17. Reams AB, Roth JR. Mechanisms of gene duplication and amplification. *Cold Spring Harb Perspect Biol.* 2015;7(2): 1–27.
18. Tautz D, Domazet-Lošo T. The evolutionary origin of orphan genes. *Nat Rev Genet.* 2011;12(10):692–702.
19. West M. Bayesian factor regression models in the large *p*, small *n* paradigm. *Bayesian Stat.* 2003;7(2003): 723–732.
20. De Jong H. Modelling and simulation of genetic regulatory systems: a literature review. *J Comput Biol.* 2002;9: 67–103.
21. Bell G. *Selection: The Mechanism of Evolution.* Chapman & Hall; 1997:699.
22. Bermingham ML, Pong-Wong R, Spiliopoulou A, Hayward C, Rudan I, Campbell H, Wright AF, Wilson JF, Agakov F, Navarro P, Haley CS. Application of high-dimensional feature selection: evaluation for genomic prediction in man. *Sci Rep.* 2015;5. Art. No. 10312.
23. Ustunkar G, Ozogur-Akyuz S, Weber GW, Friedrich CM, Aydin Son Yesim. Selection of representative SNP sets for genome-wide association studies: a meta-heuristic approach. *Optim Lett.* 2011;6(6):1207–1218.
24. Shah SC, Kusiak A. Data mining and genetic algorithm based gene/SNP selection. *Artif Intell Med.* 2004;31(3): 183–196.
25. Meiri R, Zahavi J. Using simulated annealing to optimize the feature selection problem in marketing applications. *Eur J Oper Res.* 2006;171(3):842–858.
26. Duval B, Hao JK, Hernandez JC. A memetic algorithm for gene selection and molecular classification of an cancer. In: *Proc. Of the 11th Annual Conference on Genetic and Evolutionary Computation.* 2009:201–208.
27. Jirapech-Umpai T, Aitken S. Feature selection and classification for microarray data analysis: evolutionary methods for identifying predictive genes. *BMC Bioinforma.* 2005; 6(1):1–11.
28. Pahikkala T, Tsivtsivadze E, Airola A, Järvinen J, Boberg J. An efficient algorithm for learning to rank from preference graphs. *Mach Learn.* 2009;75(1):129–165.
29. Emmert-Streib F, Glazko GV, Altay G, Matos Simoes R. Statistical inference and reverse engineering of gene regulatory networks from observational expression data. *Front Genet.* 2012;3(8):1–15.
30. Ironi L, Tran DX. Nonlinear and temporal multiscale dynamics of gene regulatory networks: a qualitative simulator. *Math Comput Simul.* 2016;125:15–37.

31. Chena Y, Mazlackb LJ, Minaib AA, Lua LJ. Inferring causal networks using fuzzy cognitive maps and evolutionary algorithms with application to gene regulatory network reconstruction. *Appl Soft Comput.* 2015;37:667−679.

32. Montes RAC, Herrera-Ubaldo H, Serwatowska J, de Folter S. Towards a comprehensive and dynamic gynoecium gene regulatory network. *Curr Plant Biol.* 2015; 3−4:3−12.

33. Patel P, Mandlik V, Singh S. LmSmdB: an integrated database for metabolic and gene regulatory network in Leishmania major and Schistosoma mansoni. *Genomics Data.* 2016;7:115−118.

34. Chowdhury AR, Chetty M. Network decomposition based large-scale reverse engineering of gene regulatory network. *Neurocomputing.* 2015;160:213−227.

35. Ito S, Ichinose T, Shimakawa M, Izumi N, Hagihara S, Yonezaki N. Qualitative analysis of gene regulatory networks by temporal logic. *Theor Computer Sci.* 2015;594: 151−179.

36. Hecker M, Lambeck S, Toepfer S, Someren E, Guthke R. Gene regulatory network inference: data integration in dynamic models-a review. *Biosystems.* 2009;96(1): 86−103.

37. Politano G, Savino A, Benso A, Carlo SD, Rehman HU, Vasciaveo A. Using Boolean networks to model post-transcriptional regulation in gene regulatory networks. *J Comput Sci.* 2014;5:332−344.

38. Chueh TH, Shing Lu HH. Inference of biological pathway from gene expression profiles by time delay boolean networks. *PLoS ONE.* 2012;7(8):1−8.

39. Kim H, Lee JK, Park T. Boolean networks using the chi-square test for inferring large-scale gene regulatory networks. *BMC Bioinforma.* 2007;8(37):1−15.

40. Silvescu A, Honavar V. Temporal boolean network models of genetic networks and their inference from gene expression time series. *Complex Syst.* 2001;13(1): 1−6.

41. Liu G, Feng W, Wang H, Liu L, Zhou C. Reconstruction of gene regulatory networks based on two-stage Bayesian network structure learning algorithm. *J Bionic Eng.* 2009; 6:86−92.

42. Adabor ES, Acquaah-Mensah GK, Oduro FT. SAGA: a hybrid search algorithm for Bayesian network structure learning of transcriptional regulatory networks. *J Biomed Inf.* 2014;53:27−35.

43. Van der Heijden M, Velikova M, Lucas PJF. Learning Bayesian networks for clinical time series analysis. *J Biomed Inf.* 2014;48:94−105.

44. Wang M, Chen Z, Cloutier S. A hybrid Bayesian network learning method for constructing gene networks. *Comput Biol Chem.* 2007;31:361−372.

45. Labatut V, Pastor J, Ruff S, Démonet J, Celsis P. Cerebral modelling and dynamic Bayesian networks. *Artif Intell Med.* 2004;30:119−139.

46. Kim S, Imoto S, Miyano S. Dynamic Bayesian network and nonparametric regression for nonlinear modelling of gene networks from time series gene expression data. *BioSystems.* 2004;75:57−65.

47. Mary MJ, Deecaraman M, Vijayalaskshmi M, Umashankar V. A systemic review of differential regulation of genes in polycystic ovarian syndrome disease. *Int J Pharma Bio Sci.* 2015;6(2):893−900.

48. Chen R, Resnick SM, Davatzikos C, Herskovits EH. Dynamic Bayesian network modelling for longitudinal brain morphometry. *NeuroImage.* 2012;59:2330−2338.

49. Peña JM, Björkegren J, Tegnér J. Learning dynamic Bayesian network models via cross-validation. *Pattern Recognit Lett.* 2005;26:2295−2308.

50. Andrusiewicza M, Słowikowskia B, Skibinskaa I, Wołun-Cholewaa M, Dera-Szymanowska A. Selection of reliable reference genes in eutopic and ectopic endometrium for quantitative expression studies. *Biomed Pharmacother.* 2016;78:66−73.

51. Mohammadi M, Noghabi HS, Hodtani GA, Mashhadi HR. Robust and stable gene selection via maximum−minimum correntropy criterion. *Genomics.* 2016;107:83−87.

52. Aguas R, Ferguson NM, Pond SLK. Feature selection methods for identifying genetic determinants of host species in RNA viruses. *PLoS Comput Biol.* 2013;9(10): 1−10.

53. ManChon U, Talevich E, Katiyar S, Rasheed K, Kannan N. Prediction and prioritization of rare oncogenic mutations in the cancer kinome using novel features and multiple classifiers. *PLoS Comput Biol.* 2014;10(4):1−12.

54. Fuentes A, Ortiz J, Saavedra N, Salazar LA, Meneses C, Arriagada C. Reference gene selection for quantitative real-time PCR in solanum lycopersicum L. inoculated with the mycorrhizal fungus Rhizophagus irregularis. *Plant Physiol Biochem.* 2016;101:124−131.

55. Diaz-Uriate R, de Andres SA. Gene selection and classification of microarray data using random forest. *BMC Bioinforma.* 2006;7(3):1−13.

56. Shreem SS, Abdullah S, Nazri MZA. Hybridizing harmony search with a Markov blanket for gene selection problems. *Inf Sci.* 2014;258:108−121.

57. Cai H, Ruan P, Ng M, Akutsu T. Feature weight estimation for gene selection: a local hyperlinear learning approach. *BMC Bioinforma.* 2014;15:1−13.

58. Han F, Sun W, Ling QH. A Novel strategy for gene selection of microarray data based on Gene-to-Class sensitivity information. *PLoS ONE.* 2014;9(5):1−17.

59. Model F, Adorjan P, Olek A, Piepenbrock C. Feature selection for DNA methylation based cancer classification. *Bioinformatics.* 2001;1(17):157−164.

60. Li T, Zhang C, Ogihara M. A comparative study of feature selection and multiclass classification methods for tissue classification based on gene expression. *Bioinformatics.* 2004;20(15):2429−2437.

61. Mundra PA, Rajapakse JC. Gene and sample selection for cancer classification with support vectors based t-statistic. *Neurocomputing.* 2010;73:2353−2362.

62. Kira K, Rendell LA. A Feature selection problem: traditional methods and a new algorithm. In: *Proc. Of the 10th National Conference on Artificial Intelligence.* 1992: 129−134.

63. Pechenizkiy M, Puuronen S, Tsymbal A. The impact of sample reduction on PCA-based feature extraction for supervised learning. In: *Proc. Of the 21st ACM Symposium on Applied Computing.* 2006:553−558.

64. Cavill R, Keun H, Holmes E, Lindon J, Nicholson J, Ebbels T. Genetic algorithms for simultaneous variable and sample selection in metabonomics. *Bioinformatics.* 2009;25(1):112−118.

65. Cawley GC, Talbot NLC. Gene selection in cancer classification using sparse logistic regression with Bayesian regularization. *Bioinformatics.* 2006;22(19):2348−2355.

66. Fitzgerald JD, Rowekamp RJ, Sincich LC, Sharpee TO. Second order dimensionality reduction using minimum and maximum mutual information models. *PLoS ONE.* 2011;7(11):1−9.

67. Piao Y, Piao M, Park K, Ryu KH. An ensemble correlation-based gene selection algorithm for cancer classification with gene expression data. *Bioinformatics.* 2012;28(24):3306−3315.

68. Han F, Yang C, Wu Y, et al. A gene selection method for microarray data based on binary PSO encoding gene-to-class sensitivity information. *IEEE Trans Comput Biol Bioinform.* 2015:1−13.

69. Meng J, Zhang J, Luan Y. Gene selection integrated with biological knowledge for plant stress response using neighborhood system and rough set theory. *IEEE/ACM Trans Comput Biol Bioinform.* 2015;12(2):433−444.

70. Liao B, Jiang Y, Liang W, Zhu W, Cai L, Cao Z. Gene selection using locality sensitive laplacian score. *IEEE/ACM Trans Comput Biol Bioinform.* 2014;11(6):1146−1155.

71. Pang H, George SL, Hui K, Tong T. Gene selection using iterative feature elimination random forests for survival outcomes. *IEEE/ACM Trans Comput Biol Bioinform.* 2012;9(5):1422−1431.

72. Ji G, Yang Z, You W. PLS-based gene selection and identification of tumor-specific genes. *IEEE Trans Syst Man, Cybern—Part C Appl Rev.* 2011;41(6):830−841.

73. Li J, Su H, Chen H, Futscher BW. Optimal search-based gene subset selection for gene array cancer classification. *IEEE Trans Inf Technol Biomed.* 2007;11(4):398−405.

74. Bontempi G. A blocking strategy to improve gene selection for classification of gene expression data. *IEEE/ACM Trans Comput Biol Bioinform.* 2007;4(2):293−300.

75. Au W, Chan KCC, Wong AKC, Wang Y. Attribute clustering for Grouping, selection, and classification of gene expression data. *IEEE/ACM Trans Comput Biol Bioinform.* 2005;2(2):83−101.

76. Guyon I, Weston J, Barnhill S, Vapnik V. Gene selectin for cancer classification using support vector machine. *Mach Learn.* 2002;46:389−422.

77. Li X, Peng S, Chen J, Li B, Zhang H, Lai M. SVM-T-RFE: a novel gene selection algorithm for identifying metastasis-related genes in colorectal cancer using gene expression profiles. *Biochem Biophysical Res Commun.* 2012;419:148−153.

78. Hidalgo-Muñoz AR, López MM, Santos IM, et al. Application of SVM-RFE on EEG signals for detecting the most relevant scalp regions linked to affective valence processing. *Expert Syst Appl.* 2013;40:2102−2108.

79. Spetale FE, Bulacio P, Guillaume S, Murillo J, Tapia E. A spectral envelope approach towards effective SVM-RFE on infrared data. *Pattern Recognit Lett.* 2016;71:59−65.

80. Shieh M, Yang CC. Multiclass SVM-RFE for product form feature selection. *Expert Syst Appl.* 2008;35:531−541.

81. Huang M, Hung YH, Lee WM, Li RK, Jiang BR. SVM-RFE based feature selection and Taguchi parameters optimization for multiclass SVM classifier. *Hindawi, Sci World J.* 2014;2014:1−10.

82. Tang Y, Zhang YQ, Huang Z. FCM-SVM-RFE gene feature selection algorithm for Leukemia classification from microarray gene expression data. In: *Proc. of the 2005 IEEE International Conference on Fuzzy Systems.* 2005:95−101.

83. Duan KB, Rajapakse JC, Wang H, Azuaje F. Multiple SVM-RFE for gene selection in cancer classification with expression data. *IEEE Trans Nanobioscience.* 2005;4(3):228−234.

84. Srinivasan U, Chen P, Diao Q, et al. Characterization and analysis of HMMER and SVM-RFE parallel bioinformatics applications. In: *Proc. of the IEEE Workload Characterization Symposium.* 2005:87−98.

85. Yuan Y, Hrebien L, Kam M. Speed up SVM-RFE Procedure using Margin Distribution. In: *Proc. of the IEEE Workshop on Machine Learning for Signal Processing.* 2005:297−302.

86. Tang Y, Zhang YQ, Huang Z, Hu X. Granular SVM-RFE gene selection algorithm for reliable prostate cancer classification on microarray expression data. In: *Proc. of the 5th IEEE Symposium on Bioinformatics and Bioengineering.* 2005:1−4.

87. Yoon S, Kim S. Multiple SVM-RFE using boosting for mammogram classification. In: *Proc. of the International Joint Conference on Computational Sciences and Optimization.* 2009:740−742.

88. Wang L, Pei Y, Chen J, Zhao X, Cui H, Cui H. Feature selection and prediction of sub-health state using SVM-RFE. In: *Proc. of the International Conference on Artificial Intelligence and Computational Intelligence.* 2010:199−202.

89. Wang J, Shan G, Duan X, Wen B. Improved SVM-RFE feature selection method for multi-SVM classifier. In: *Proc. of the International Conference on Electrical and Control Engineering.* 2011:1592−1595.

90. Li X, Shao Q, Wang J. Improved automatic filtering algorithm for imbalanced classification based on SVM-RFE. In: *Proc. of the IEEE International Conference on Bioinformatics and Biomedicine.* 2013:110−113.

91. Zhang J, Hu X, Li P, He W, Li H. A hybrid feature selection approach by correlation based filters and SVM-RFE. In: *Proc. of the 22nd International Conference on Pattern Recognition.* 2014:3684−3689.

92. Yin J, Hou J, She Z, Yang C, Yu H. Improving the performance of SVM-RFE on classification of pancreatic cancer data. In: *Proc. of the IEEE International Conference on Industrial Technology*. 2016:956−961.

93. Yoon S, Kim S. Mutual information-based SVM-RFE for diagnostic classification of digitized mammograms. *Pattern Recognit Lett*. 2009;30:1489−1495.

94. Sha-Sha W, Hui-Juan L, Wei J, Chao L. A construction method of gene expression data based on information gain and extreme learning machine classifier on cloud platform. *Int J Database Theor Appl*. 2014;7(2):99−108.

95. Lei S. A feature selection method based on information gain and genetic algorithm. In: *Proc. of the International Conference on Computer Science and Electronics Engineering*. 2012:355−358.

96. Wei-qiang L, Xiao-Feng W. Improved method of feature selection based on information gain. In: *Proc. of the Spring Congress on Engineering and Technology*. 2012:1−4.

97. Sehhati M, Mehridehnavi A, Rabbani H, Pourhossein M. Stable gene signature selection for prediction of breast cancer recurrence using joint mutual information. *IEEE/ACM Trans Comput Biol Bioinform*. 2015;12(6): 1440−1448.

98. Wu G, Xu J. Optimized approach of feature selection based on information gain. In: *Proc. of 2015 International Conference on Computer Science and Mechanical Automation (CSMA)*. 2015:157−161.

99. Shaltout N, Moustafa M, Rafea A, Moustafa A, ElHefnawi M. Comparing PCA to information gain as a feature selection method for influenza-a classification. In: *Proc. of the International Conference on Intelligent Informatics and Biomedical Sciences*. 2015:279−283.

100. Azhagusundari B, Thanamani AS. Feature selection based on information gain. *Int J Innovat Technol Explor Eng*. 2013;2(2):18−21.

101. Correa NM, Li Y, Adali T, Calhoun VD. Fusion of fMRI, sMRI, and EEG data using canonical correlation analysis. In: *Proc. of the IEEE International Conference on Acoustics, Speech, and Signal Processing*. 2009:385−388.

102. Yan J, Zheng W, Zhou X, Zhao Z. Sparse 2-D canonical correlation analysis via low rank matrix approximation for feature extraction. *IEEE Signal Process Lett*. 2012; 19(1):51−54.

103. Li S, Xu G, Feng Y. Gaussian Bayesian network structure learning strategies based on canonical correlation analysis. In: *Proc. of the IEEE International Conference on Mechatronics and Automation*. 2012:156−161.

104. Zhang Z, Zhao M, Chow TWS. Binary and multi-class group sparse canonical correlation analysis for feature extraction and classification. *IEEE Trans Knowl Data Eng*. 2013;25(10):2192−2205.

105. Gao L, Qi L, Chen E, Guan L. Discriminative multiple canonical correlation analysis for multi-feature information fusion. In: *Proc. of the IEEE International Symposium on Multimedia*. 2012:36−43.

106. Zu C, Zhang D. Canonical sparse cross-view correlation analysis. *Neurocomputing*. 2016;191:263−272.

107. Wang S, Lu J, Gu X, Shen C, Xia R, Yang J. Canonical principal angles correlation analysis for two-view data. *J Vis Commun Image Represent*. 2016;35:209−219.

108. Xing X, Wang K, Yan T, Lv Z. Complete canonical correlation analysis with application to multi-view gait recognition. *Pattern Recognit*. 2016;50:107−117.

109. Zhai D, Zhang Y, Yeung D, Chang H, Chen X, Gao W. Instance-specific canonical correlation analysis. *Neurocomputing*. 2015;155:205−218.

110. Zhang LH. Uncorrelated trace ratio linear discriminant analysis for under sampled problems. *Pattern Recognit Lett*. 2011;32:476−484.

111. Zhao M, Zhang Z, Chow TWS. Trace ratio criterion based generalized discriminative learning for semi-supervised dimensionality reduction. *Pattern Recognit*. 2012;45: 1482−1499.

112. Liu Y, Nie F, Wu J, Chen L. Efficient semi-supervised feature selection with noise insensitive trace ratio criterion. *Neurocomputing*. 2013;105:12−18.

113. Liu M, Sun W, Liu B. Multiple kernel dimensionality reduction via spectral regression and trace ratio maximization. *Knowl-Based Syst*. 2012;83:159−169.

114. Nie F, Xiang S, Jia Y, Zhang C, Yan S. Trace ratio criterion for feature selection. In: *Proc. of the Twenty-third AAAI Conference on Artificial Intelligence*. 2008:671−676.

115. Zhao M, Zhang Z, Chow TWS. Itr-score algorithm: an efficient trace ratio criterion based algorithm for supervised dimensionality reduction. In: *Proc. of the International Joint Conference on Neural Networks*. 2011: 145−152.

116. Huang Y, Xu D, Nie F. Semi-supervised dimension reduction using trace ratio criterion. *IEEE Trans On Neural Netw And Learn Syst*. 2012;23(3):519−526.

117. Zhao M, Chan RHM, Tang P, Chow TWS, Wong SWH. Trace ratio linear discriminant analysis for medical diagnosis: a case study of dementia. *IEEE Signal Process Lett*. 2013;20(5):431−434.

118. Li C, Shi C, Zhang H, Hui C, Lam K, Zhang S. Cost-sensitive feature selection in medical data analysis with trace ratio criterion. In: *Proc. of the International Conference on Signal Processing*. 2014:1077−1082.

119. Gao D, Li M, Li J, et al. Effects of various typical electrodes and electrode gels combinations on MRI signal-to-noise ratio and safety issues in EEG-fMRI recording. *Biocyberntics Biomed Eng*. 2016;36:9−18.

120. Morawski RZ, Miekina A. Application of principal components analysis and signal-to-noise ratio for calibration of spectrophotometric analysers of food. *Measurement*. 2016;79:302−310.

121. Qian Y, Shen M, Amoureux JP, Noda I, Hu B. The dependence of signal-to-noise ratio on number of scans in covariance spectroscopy. *Solid State Nucl Magn Reson*. 2014;59:31−33.

122. Venet D, Detours V, Bersini H. A measure of the signal-to-noise ratio of microarray samples and studies using gene correlations. *PLOS One.* 2012;7(2):1–9.

123. Hengpraprohm S, Chongstitvatana P. Feature selection by weighted-SNR for cancer microarray data classification. *Int J Innovat Comput Inf Control.* 2009; 5(12):4627–4635.

124. Goh L, Song Q, Kasabov A. A novel feature selection method to Improve classification of gene expression data. *Proc 2nd Asia-Pacific Bioinformatics Conf.* 2004;29: 161–166.

125. Aziz R, Verma CK, Srivastava N. A weighted-SNR feature selection from independent component subspace for NB classification of microarray data. *Int J Adv Biotechnol Res.* 2015;6(2):245–255.

126. Kourid A. Iterative MapReduce for feature selection. *Int J Eng Res Technol.* 2014;3(7):1788–1793.

127. Maulik U, Chakraborty D. Fuzzy preference based feature selection and semi-supervised SVM for cancer classification. *IEEE Trans On Nano-bioscience.* 2014; 13(2):152–160.

128. Kitano H. System biology: a brief overview. *Science.* 2002; 295(5560):1662–1664.

129. Babu MM, Teichmann SA. Evolution of transcription factors and the gene regulatory network in *Escherichia coli. Nucleic Acids Res.* 2013;31:1234–1244.

130. Gomaa WE. Modeling gene regulatory networks: a survey in the Egypt. In: *Proc. 9th IEEE/ACS International Conference on Computer Systems and Applications (AICCSA).* 2011:204–208.

131. Schlitt T, Brazma A. Modelling gene networks at different organizational levels. *FEBS Lett.* 2005;579(8):859–866.

132. Schlitt T, Brazma A. Current approaches to gene regulatory network modeling. *BMC Bioinforma.* 2007;8(6): 1–22.

133. Tyagi V, Mishra A. A survey on different feature selection methods for microarray data analysis. *Int J Computer Appl.* 2013;67(16):36–40.

134. Alshamlan HM, Badr GH, Alohali YA. The performance of bio-inspired evolutionary gene selection methods for cancer classification using microarray dataset. *Int J Biosci Bioinform.* 2014;4(3):166–170.

135. Lazar C, Taminau J, Meganck S, et al. A Survey on filter techniques for feature selection in Gene Expression Microarray analysis. *IEEE/ACM Trans Comput Biol Bioinforma (TCBB).* 2012;9(4):1106–1119.

136. Abu Shanab A, Khoshgoftaar TM, Wald R. Evaluation of wrapper-based feature selection using hard, moderate, and easy bioinformatics data. In: *Proc. of the IEEE International Conference on Bioinformatics and Bioengineering (BIBE).* 2014:149–155.

137. Maldonado S, Weber R, Famili F. Feature selection for high dimensional class-imbalanced datasets using support vector machines. *Inf Sci.* 2014;286:228–246.

138. Cateni S, Colla V, Vannucci M. A Hybrid feature selection method for classification purposes. In: *Proc. of the European Modelling Symposium (EMS).* 2014:39–44.

139. Srivastava B, Srivastava R, Jangid M. Filter vs wrapper approach for optimum gene selection of high dimensional gene expression dataset: an analysis with cancer datasets. In: *Proc. of the International Conference on High Performance Computing and Applications.* 2014:1–6.

140. Phuong TM, Lin Z, Altman RB. Choosing SNPs using feature selection. In: *Proc. of IEEE Computational Systems Bioinformatics Conference.* 2005:30–39.

141. Blum AL, Langley P. Selection of relevant features and examples in machine learning. *Artif Intell.* 1997; 97(1–2):245–270.

142. Wang X, Gotoh O. A Robust Gene Selection Method for microarray based cancer classification. *Cancer Inf.* 2010; 9:15–30.

143. Zhou N, Wang L. A Modified T-test feature selection method and its application on the HapMap Genotype Data. *Genomics Proteomics Bioinformatics.* 2007;5(3):242–249.

144. Baldi P, Long AD. A Bayesian framework for the analysis of microarray expression data: regularized t-test and statistical inferences of gene changes. *Bioinformatics.* 2001; 17(6):509–519.

145. Spokoiny V, Dickhaus T. *Bayes Estimation, Basics of Modern Mathematical Statistics.* 2014:173–194.

146. Gene Expression Omnibus (GEO), GSE8671Series: https://www.ncbi.nlm.nih.gov/geo/query/acc.cgi?acc=GSE8671, GSE8671 series.

147. Leukemia Set, http://www.github.com/Leukemia.gct.

148. Broad institute, http://www.broadinstitute.org/cgi-bin/cancer/datasets.cgi.

149. Shipp MA, Ross KN, Jackson DG, et al. Diffuse large B-cell lymphoma outcome prediction by gene-expression profiling and supervised machine learning. *Nat Med.* 2002;8:68–74.

150. Singh D, Febbo PG, Ross K, et al. Gene expression correlates of clinical prostate cancer behaviour. *Cancer Cell.* 2002;1:203–209.

151. Chapelle O, Vapnik V, Bousquet O, Mukherjee S. Choosing multiple parameters for support vector machines. *Mach Learn.* 2002;46:131–159.

152. Mishra S, Mishra D. SVM-BT-RFE: an improved gene selection framework using Bayesian T-test embedded in support vector machine (recursive feature elimination) algorithm. *Karbala Int J Mod Sci.* 2015;1(2):86–96.

153. Leitner F, Krallinger M, Tripathi SL, Kuiper M, Lægreid A, Valencia A. Mining cis-regulatory transcription networks from literature. In: *Proc. of BioLINK Special Interest Group (ISBM/ECCB).* 2013:5–12.

154. Alon U, Barkai N, Notterman DA, et al. Broad patterns of gene expression revealed by clustering analysis of tumor and normal colon tissues probed by oligonucleotide arrays. *Proc Natl Acad Sci USA.* 1999;96(12):6745–6750.

155. Nowozin S. Improved Information Gain estimates for decision tree induction. In: *Proceedings of the 29th International Conference on Machine Learning.* 2012:1–8.

156. Xu JDQ. Attribute selection based on information gain ratio in fuzzy rough set theory with application to tumor classification. *Appl Soft Comput.* 2013;13(1):211–221.

157. Shang C, Li M, Feng S, Jiang Q, Fan J. Feature selection via maximizing global information gain for text classification. *Knowl-Based Syst*. 2013;54:298−309.

158. Frieden BR, Gatenby RA. Principle of maximum fisher information from hardy's axioms applied to statistical systems. *Phys Rev E*. 2013;88(4):1−13.

159. Zhu L, Miao L, Zhang D. Iterative Laplacian score for feature selection, pattern recognition. *Ser Commun Computer Inf Sci*. 2012;321:80−87.

160. Wang H, Yan S, Xu D, Tang X, Huang T. Trace ratio vs. ratio trace for dimensionality reduction. In: *Proc. of 2007 IEEE Conference on Computer Vision and Pattern Recognition*. 2007:1−8.

161. Klami A, Viratanen S, Kaski S. Bayesian canonical correlation analysis. *J Mach Learn Res*. 2013;14:965−1003, 966−1003.

162. Wang S, Lu J, Gu X, Weyori BA, Yang J. Unsupervised discriminant canonical correlation analysis based on spectral clustering. *Neurocomputing*. 2016;171:425−433.

163. Tenenhaus A, Philippe C, Frouin V. Kernel generalized canonical correlation analysis. *Comput Stat Data Anal*. 2015;90:114−131.

164. Kuncheva LI. A stability index for feature selection. In: *Proc. of 25th IASTED Int. Conf. on Artificial Intelligence and Applications*. 2007:390−395.

165. Lauwerys BR, Hernández-Lobato D, Gramme P, et al. Heterogeneity of synovial molecular patterns in patients with arthritis. *PLoS ONE*. 2015:1−18.

166. Pai DR, Lawrence KD, Klimberg RK, Lawrence SM. Analyzing the balancing of error rates for multi-group classification. *Expert Syst Appl*. 2012;39(17): 12869−12875.

167. Pearson K. Notes on regression and inheritance in the case of two parents. *Proc R Soc Lond*. 1895;58:240−242.

168. Fisher RA. On the probable error of a coefficient of correlation deduced from a small sample. *Metron*. 1921;1(4): 3−32. Retrieved in 2009.

169. Shannon P, Markiel A, Ozier O, et al. Cytoscape: a software environment for integrated models of biomolecular interaction networks. *Genome Res*. 2003;13(11): 2498−2504.

170. Saito R, Smoot ME, Ono K, et al. A travel guide to cytoscape plugins. *Nat Methods*. 2012;9(11):1069−1076.

Index

Note: Page numbers followed by "f" indicate figures, "t" indicate tables.

123

Printed in the United States
By Bookmasters